ELECTRICAL MOTOR CONTROLS

WORKBOOK

4

AMERICAN TECHNICAL PUBLISHERS, INC.
HOMEWOOD, ILLINOIS 60430

Gary Rockis
Glen A. Mazur

Ryan Skinner

1 2 3 4 5 6 7 8 9 – 97 – 9 8 7 6 5 4 3

Printed in the United States of America

ISBN 0-8269-1672-4

CONTENTS

AC/DC Magnetic Contactors and Motor Starters

Time Delay and Logic

Control Devices

Reversing Motor Circuits

Power Distribution Systems

Solid-State Electronic Control Devices

Electromechanical and Solid-State Relays

Photoelectric and Proximity Controls

14 Programmable Controllers

15 Reduced-Voltage Starting

16 Accelerating and Decelerating Methods

17 Preventive Maintenance and Troubleshooting

Appendix

INTRODUCTION

Electrical Motor Controls Workbook is designed to reinforce the concepts, provide hands-on applications, and test the material presented in *Electrical Motor Controls*. When studying the text, pay particular attention to italicized terms, illustrations, and examples. These key elements comprise a major portion of *Electrical Motor Controls Workbook*.

Tech-Cheks

Electrical Motor Controls Workbook contains 17 Tech-Cheks. Each Tech-Chek is a series of multiple choice, completion, and matching questions that are based on the text and art in the corresponding chapter of *Electrical Motor Controls*. Always study the assigned chapter of *Electrical Motor Controls* thoroughly before completing the Tech-Cheks.

Worksheets

Electrical Motor Controls Workbook contains 111 Worksheets developed from the 17 chapters of *Electrical Motor Controls*. Worksheets provide opportunities to apply the concepts and theory in *Electrical Motor Controls* to practical problems. See Contents for a complete listing of Worksheets.

Appendix

The Appendix contains Data Sheets, charts, and tables for use with the Worksheets. See page 169 for a complete listing of Data Sheets, charts, and tables in the Appendix.

The Authors and Publisher

> ## Electrical Tools, Instruments, and Safety
>
> Name _Ryan Skinner_ Date _6-8-99_
>
> ## TECH-CHEK 1

Electrical Motor Controls

_D_ 1. Electrical tools may be organized using a(n) _____.

    A. pegboard   C. portable tool box
    B. electrician's pouch   D. A, B, and C

understood
Grounded 2. All power tools should be properly _____ before being used.

off 3. Ensure the switch is in the _____ position before connecting a tool to a power source.

hot 4. Always remove the _____ side of a fuse when removing fuses from circuits.

de energized
locked out 5. Ensure the source of electricity is open and _____ before performing any repair on a piece of electrical equipment.

_A_ 6. Class _____ fires consist of burning wood, clothing, or paper.

_C_ 7. Class _____ fires consist of burning electrical equipment.

_B_ 8. Class _____ fires consist of flammable liquids, such as gasoline and oil.

_D_ 9. Class _____ fires consist of burning combustible metals such as magnesium.

Voltmeter 10. A(n) _____ is used when checking voltage in a circuit.

clamp on Ampmeter
Ampmeter 11. A(n) _____ is used when checking the current level in a circuit.

or Multimeter

Ohm meter 12. A(n) _____ is used when checking the resistance in a component or device.

absent 13. Never assume the power is _____ when working on an electrical circuit.

_C_ 14. A(n) _____ is the best system for organizing tools to be used on the job or at a test bench.

    A. pegboard   C. portable tool box
    B. electrician's pouch   D. neither A, B, nor C

_C_ 15. Cutting tools should be _____.

    A. stainless steel   C. sharp and clean
    B. new   D. well used

_C_ 16. All power tools should be grounded unless _____.

    A. the area has low humidity   C. they are double-insulated
    B. the area is dry   D. permission has been granted to use them without grounding

_B_ **17.** A change in sound during tool operation normally indicates _____.

    A. normal operation   C. the tool is in reverse
    B. trouble of some type   D. overheating

_B_ **18.** Install a fuse _____ when replacing fuses.

    A. line side first   C. load and line side simultaneously
    B. load side first   D. with a screwdriver

_A_ **19.** Rags containing oil, gasoline, alcohol, shellac, paints, varnish, or lacquer must be _____.

    A. kept in a covered metal container   C. stored in a cool, dry place
    B. stored in a wastebasket   D. left out to dry

_B_ **20.** All unfamiliar wires should be treated as if they are _____.

    A. dead   C. harmless
    B. alive   D. A, B, and C

_A_ **21.** Acid on hands and face should be immediately washed away with plenty of _____.

    A. water   C. vaseline
    B. glycerine   D. gasoline

Information output **22.** A(n) _____ is any output device that displays data about the circuit or operation.

_B_ **23.** Lockout is the process of _____.

    A. putting a danger tag on an   C. putting a lockout tag on an
       electrical source   electrical source
    B. padlocking the electrical source   D. neither A, B, nor C

wrap-around **24.** A(n) _____ is a bar graph that displays a fraction of the full range on the graph.

analog display **25.** A(n) _____ is an electromechanical device that indicates readings on a meter by the mechanical motion of a pointer.

Analog Displays

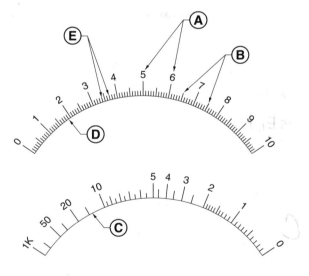

_B C_   **1.** Nonlinear scale

_B_   **2.** Secondary divisions

_E_   **3.** Subdivisions

_D_   **4.** Linear scale

_A_   **5.** Primary divisions

Electrical Tools, Instruments, and Safety

Name _____ Date _____

WORKSHEET 1-1

Tool Identification

_____ 1. Ball peen hammer

_____ 2. Chain wrench

_____ 3. Slip-joint pliers

_____ 4. Adjustable wrench

_____ 5. Phillips screwdriver

_____ 6. Flathead screwdriver

_____ 7. Locking pliers

_____ 8. Hex key wrench

_____ 9. Folding rule

_____ 10. Phillips offset screwdriver

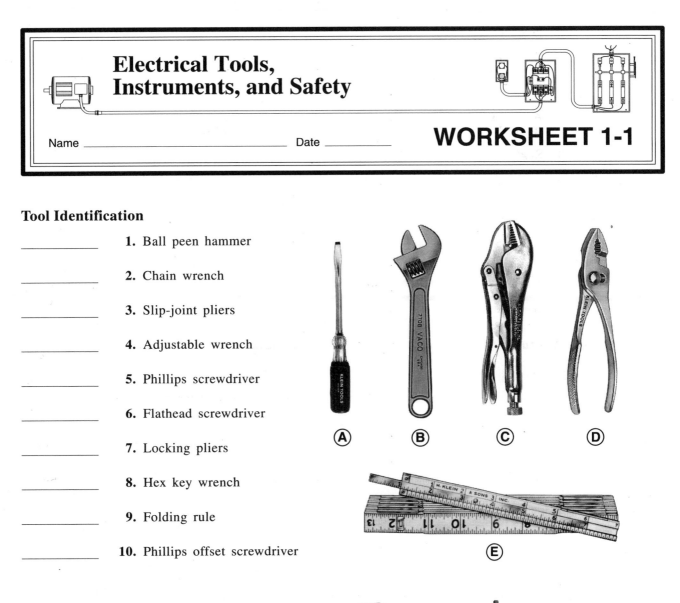

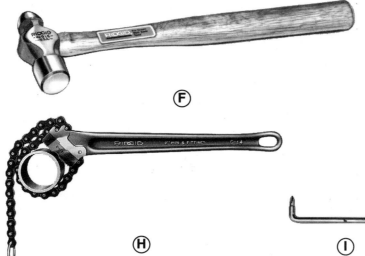

Electrical Tools, Instruments, and Safety

Name _____ Date _____

WORKSHEET 1-2

Electrical Tool Identification

_____ **1.** Cable cutter

_____ **2.** Fuse puller

_____ **3.** Rigid conduit hickey

_____ **4.** Long-nose pliers

_____ **5.** Side-cutting pliers

_____ **6.** End-cutting pliers

_____ **7.** Skinning knife

_____ **8.** Reaming tool

_____ **9.** Diagonal-cutting pliers

_____ **10.** Wire stripper

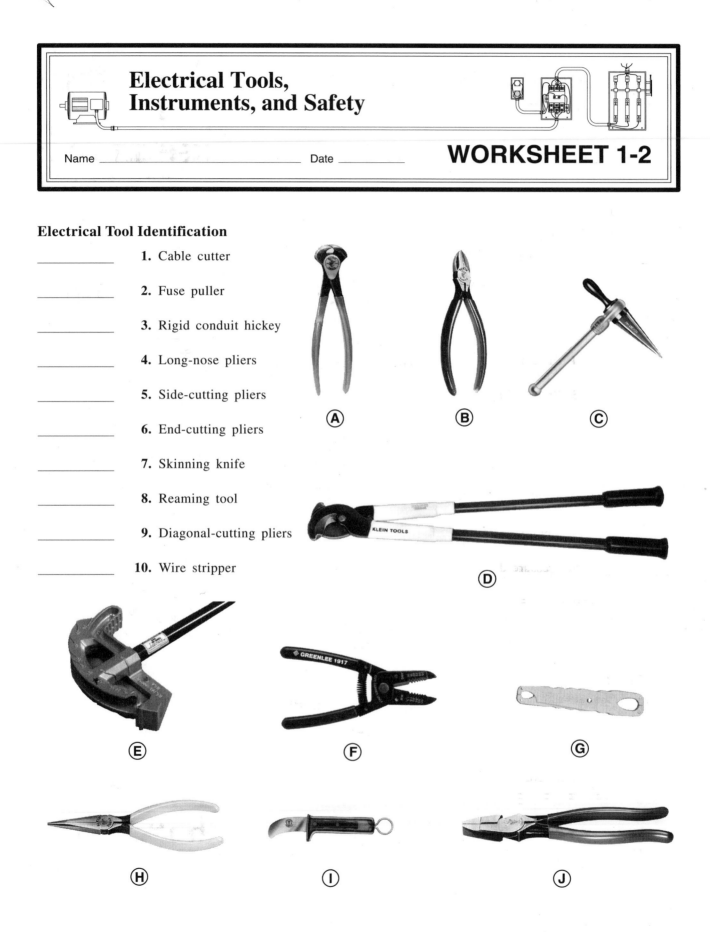

Electrical Tools, Instruments, and Safety

Name **Ryan Skinner**   Date **6-8-99**   **WORKSHEET 1-3**

Informational Outputs

D   **1.** Air velocity or force

G   **2.** Relative humidity

H   **3.** Imbalance

I   **4.** Fluid flow

C   **5.** Speed of rotating object

J   **6.** Number of devices

A   **7.** Fluid pressure

B   **8.** Heat

F   **9.** Acidity or alkalinity

E   **10.** Pressure differential

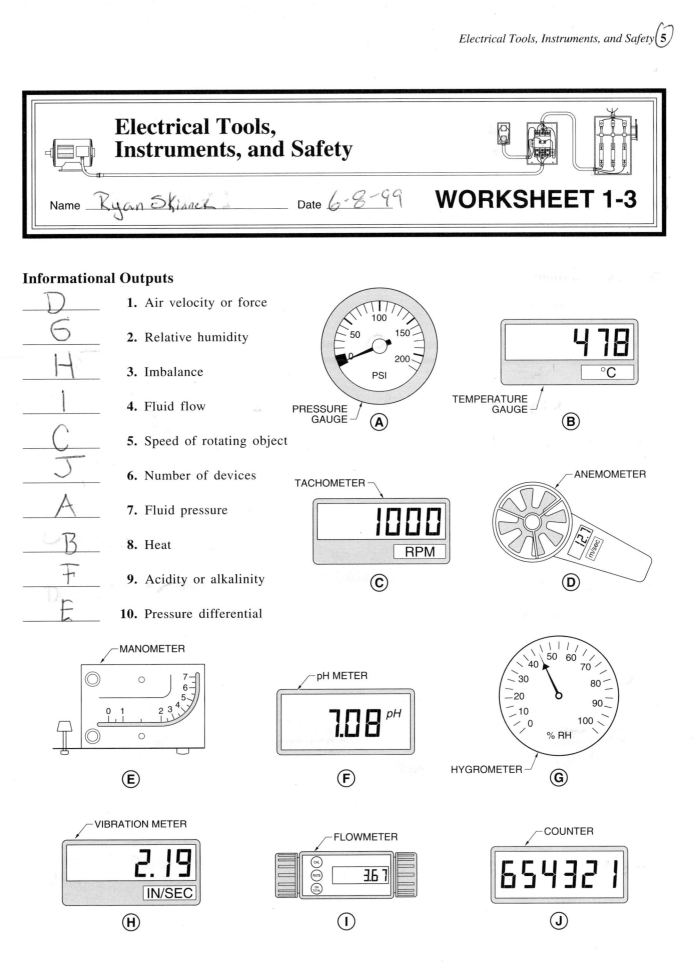

PRESSURE GAUGE — (A)

TEMPERATURE GAUGE — (B)

TACHOMETER — (C)

ANEMOMETER — (D)

MANOMETER (E)

pH METER (F)

HYGROMETER — (G)

VIBRATION METER (H)

FLOWMETER (I)

COUNTER (J)

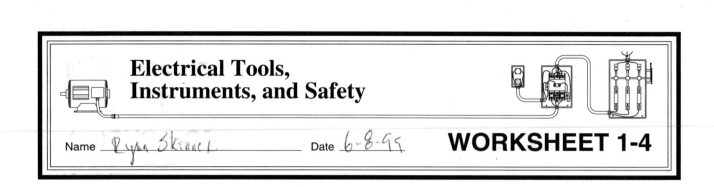

Electrical Tools, Instruments, and Safety

Name Ryan Skinner   Date 6-8-99   # WORKSHEET 1-4

Multimeter Basic Measurements

Draw the correct position of the function switches to measure the electrical quantity based on the meter connections and circuit.

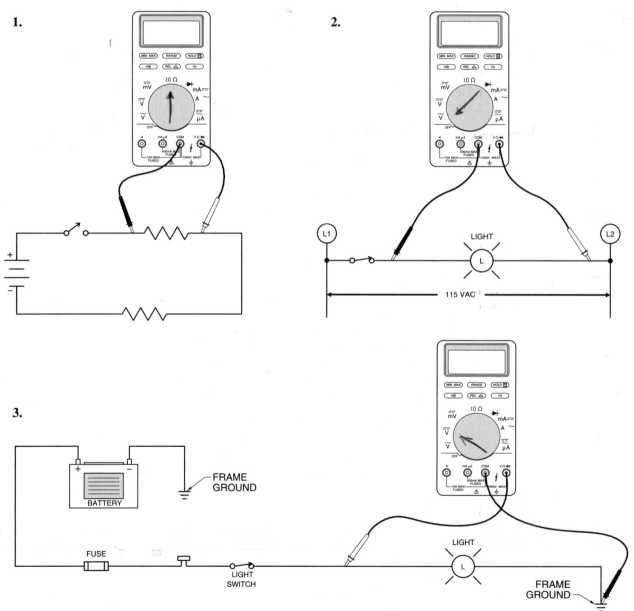

**Electrical Symbols
and Line Diagrams**

Name _____ Date _____

TECH-CHEK 2

Electrical Motor Controls

line (C) **1.** In electrical circuits, the basic means of communicating the language of control is through a _____ diagram.

A. pictorial C. line
B. schematic D. wiring

A **2.** In a line diagram, the power source is shown in _____ lines than the rest of the diagram.

A. heavier C. straighter *Its labeled L1/L2*
B. thinner D. neither A, B, nor C

B **3.** In a line diagram, the path of current flow through the various parts of the control circuit, such as the pushbuttons, etc., is shown in _____ lines than the rest of the diagram.

A. heavier C. straighter
B. thinner D. neither A, B, nor C

manual (A)
Mechanical **4.** A pushbutton is an example of a(n) _____ control switch.

A. manual C. mechanical
B. automatic D. neither A, B, nor C

(B)
Automatic **5.** A liquid level switch is an example of a(n) _____ control switch.

A. manual C. mechanical
B. automatic D. neither A, B, nor C

L2 **6.** A line diagram is always read from line 1 to line _____.

L2 **7.** In a line diagram, one side of the overload contacts are connected to line _____.

NC (B) **8.** The _____ contacts are always used when using a float switch to maintain a predetermined level.

A. NO C. auxiliary
B. NC D. neither A, B, nor C

A **9.** The _____ contacts are always used when using a float switch to control a sump pump.

A. NO C. auxiliary
B. NC D. neither A, B, nor C

Solenoid **10.** A(n) _____ is an electrical device which consists of a frame, plunger, and coil and is used to create a push or pull action.

Contactor **11.** A(n) _____ is an electrical device which consists of a frame, plunger, and coil and is used to open and close a set of contacts.

Magnetic
Motor Starter **12.** A(n) _____ is an electrical device which consists of a frame, plunger, and coil and is used to open and close a set of contacts in addition to providing overload protection.

Coil Aux memory **13.** The _____ contacts are used in the control circuit to maintain an electrical holding circuit.

Negative **14.** DC control voltage may be marked with a positive or _____ sign.

Over load **15.** For consistency, the _____ symbol is always drawn in a line diagram after the motor.

Auxiliary **16.** _____ contacts are attached to the side of a contactor and are opened or closed with the power contacts.

Heaters **17.** Overloads have _____ which sense excessive current flow to the motor.

Control **18.** _____ language is communicated using line diagrams.

Contactors **19.** Solenoids, _____, and magnetic motor starters are used for remote control of devices.

Manual **20.** A(n) _____ control circuit is a circuit that requires a person to initiate an action for the circuit to operate.

Device Identification

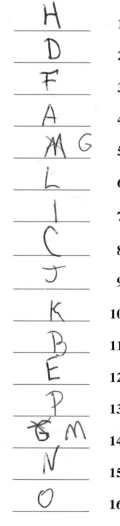

H ___ **1.** Foot switch

D ___ **2.** Silicon-controlled rectifier

F ___ **3.** NC limit switch

A ___ **4.** Pilot light

M G ___ **5.** Solenoid

L ___ **6.** Liquid level switch

I ___ **7.** 1φ motor

C ___ **8.** 3φ motor

J ___ **9.** Temperature switch

K ___ **10.** Pressure or vacuum switch

B ___ **11.** Flow switch

E ___ **12.** Control transformer

P ___ **13.** Motor starter

O M ___ **14.** Overload contacts

N ___ **15.** Electrical junction

O ___ **16.** NO pushbutton

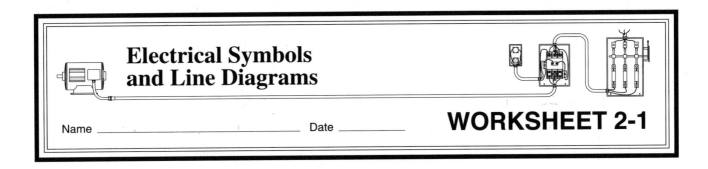

Electrical Symbols

Draw the appropriate symbol.

1. NO limit switch

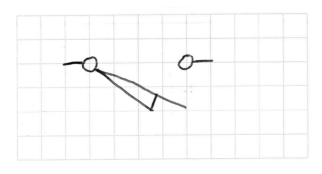

2. Circuit breaker with thermal and magnetic overload

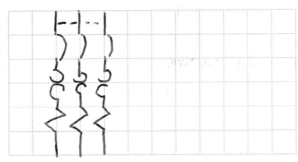

3. NO held-closed limit switch

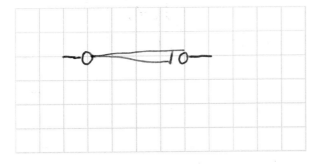

4. NO and NC pushbutton

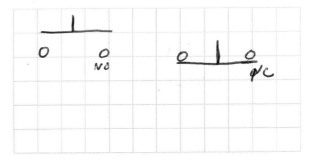

5. Single-voltage transformer

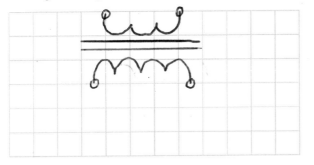

6. Dual-voltage transformer

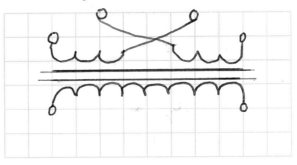

7. 3φ motor

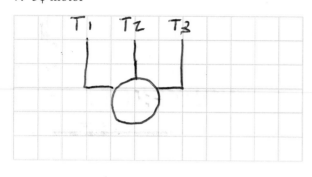

8. NO timed-closed contact

9. Disconnect

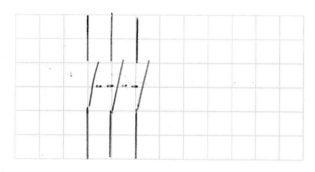

10. Two-position selector switch

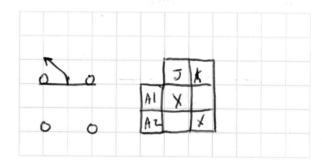

11. Red pilot light

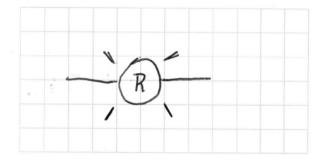

12. Photocell

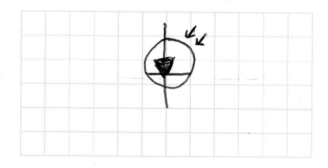

13. Full-wave rectifier

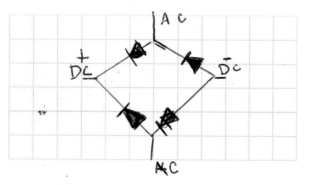

14. Shunt field

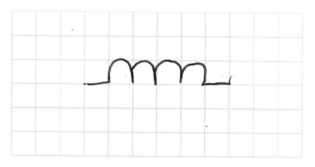

Electrical Symbols and Line Diagrams

Name _____ Date _____

WORKSHEET 2-2

Line Diagrams

Complete each line diagram with the appropriate symbol.

1. NC held-open limit switch

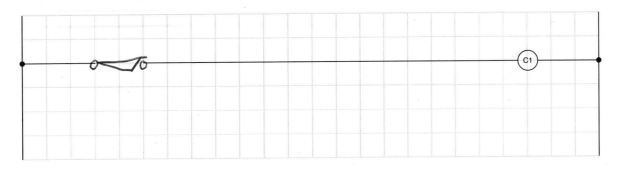

2. NC timed-closed contact

3. NC mushroom head pushbutton

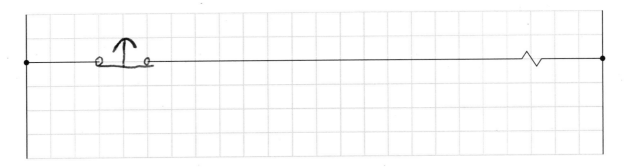

4. NO temperature-activated switch

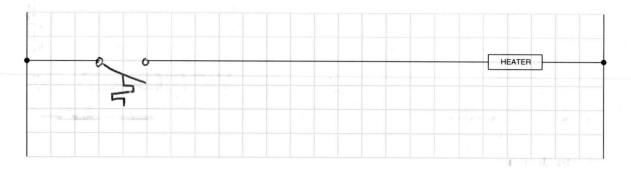

5. NO solid-state limit switch

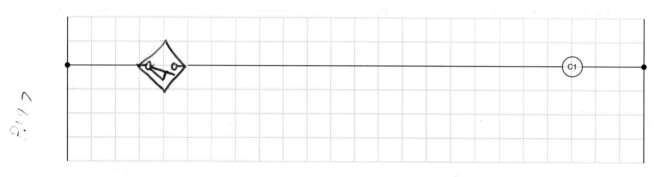

6. NC flow switch

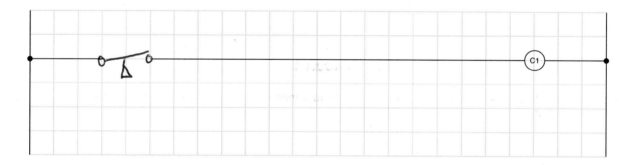

7. Thermal overload switch

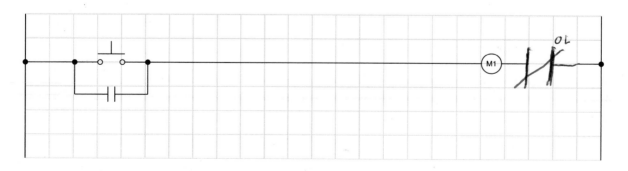

Basic Circuit Design

Complete each line diagram according to the circuit information. Use standard lettering, numbering, and coding information.

1. Design a circuit in which a NO start pushbutton controls a magnetic motor starter with three overload contacts.

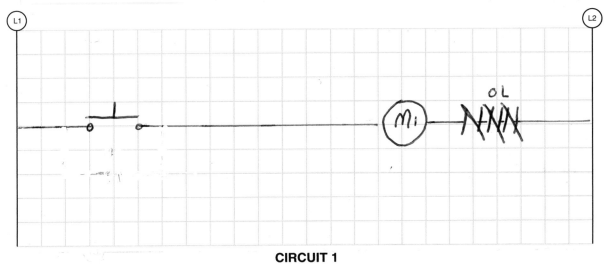

CIRCUIT 1

2. Redraw Circuit 1, adding an auxiliary contact to form a memory circuit. Add an NC stop pushbutton to turn OFF the circuit.

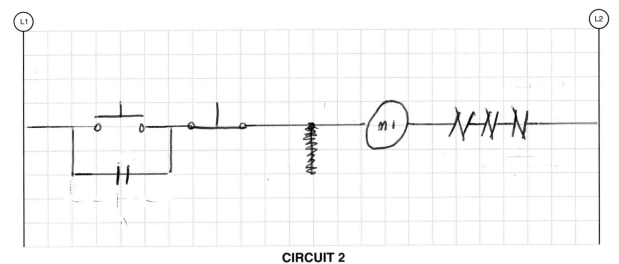

CIRCUIT 2

3. Redraw Circuit 2, adding a foot switch and a limit switch that turn OFF the motor if actuated.

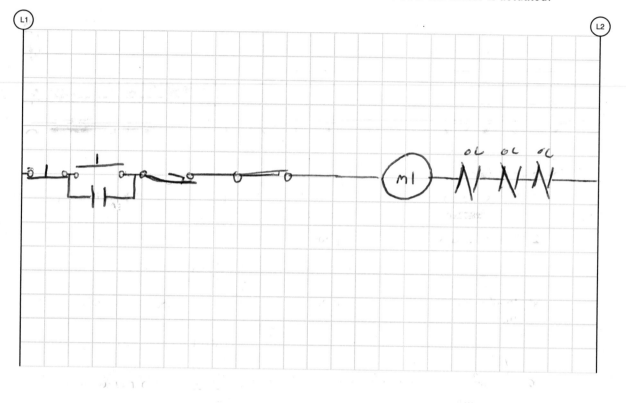

4. Design a control circuit in which a pressure switch is used to control a pump motor. The pump motor should turn ON any time the pressure drops below 30 psi.

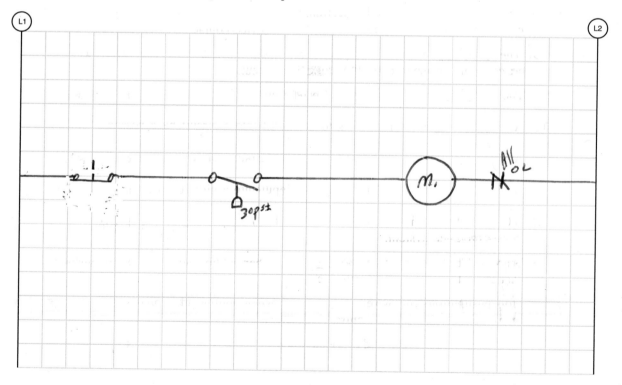

Logic Applied to Line Diagrams

Name _____ Date _____

TECH-CHEK 3

Electrical Motor Controls

Parallel   **1.** Loads are connected in _____ when more than one electrical load is connected in a line diagram.

L2   **2.** Control relays, solenoids, and pilot lights are loads connected directly to line _____.

motor starters **3.** A(n) _____ is a load that is connected indirectly to line 2.

Operating coil **4.** In a line diagram, pushbuttons, limit switches, and pressure switches are connected between line 1 and the _load_.

Numerical Cross-Reference   **5.** _____ systems help to quickly identify the location and type of contacts controlled by a given device.

number   **6.** Each wire in a control circuit is assigned a(n) _____ to keep track of the different wires that connect the components in a circuit.

action   **7.** A control circuit is composed of three basic sections, which are the signal, decision, and _____ sections.

signal   **8.** The _____ section is the section of a control circuit that starts or stops the flow of current by closing or opening the control device contacts.

decision   **9.** The _____ section is the section of a control circuit that determines what work is to be done and in what order the work is to occur.

Action   **10.** The _____ section is the section of a circuit that causes work to take place.

Manual   **11.** A(n) _____ condition refers to any input into a circuit by a person.

Mechanical   **12.** A(n) _____ condition refers to any input into a circuit by some moving part.

automatic   **13.** A(n) _____ condition refers to any input into a circuit from changes in a system.

underlined   **14.** NC contacts are indicated as a number which is _____ when applying the numerical cross-reference system.

Parallel   **15.** Start pushbuttons are wired in _____ when adding additional start pushbuttons to a standard start/stop motor control circuit.

series   **16.** Stop pushbuttons are wired in _____ when adding additional stop pushbuttons to a standard start/stop motor control circuit.

Nor   **17.** _____ circuit logic is developed when NC switches are connected in series.

Nand   **18.** _____ circuit logic is developed when NC switches are connected in parallel.

Logic Functions

B _____ 1. OR

D _____ 2. AND

E _____ 3. NOT

A _____ 4. NOR

C _____ 5. Memory

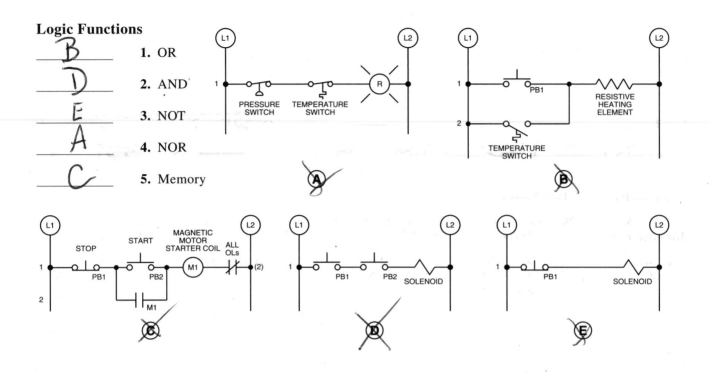

Reference Numbers

1. Add line-reference, numerical cross-reference, and wire-reference numbers per industrial standards.

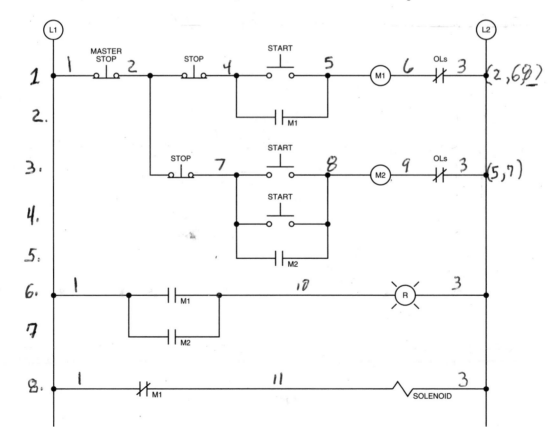

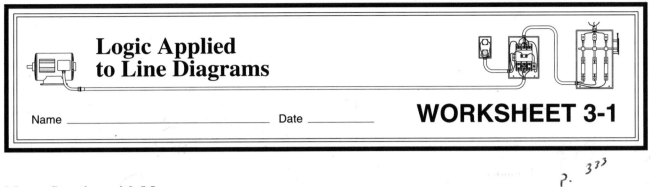

Logic Applied to Line Diagrams

Name _____ Date _____

WORKSHEET 3-1

P. 373

Motor Starting with Memory

Use standard lettering, line-reference, and cross-reference numbering systems to complete each line diagram. Wire-reference numbers are not required. Note: *The motor starter includes both NO and NC auxiliary contacts.*

1. Complete the control circuit line diagram so any one of three start pushbuttons starts the motor and any one of three stop pushbuttons stops the motor. Include memory so the motor remains running after any start pushbutton is pressed and released.

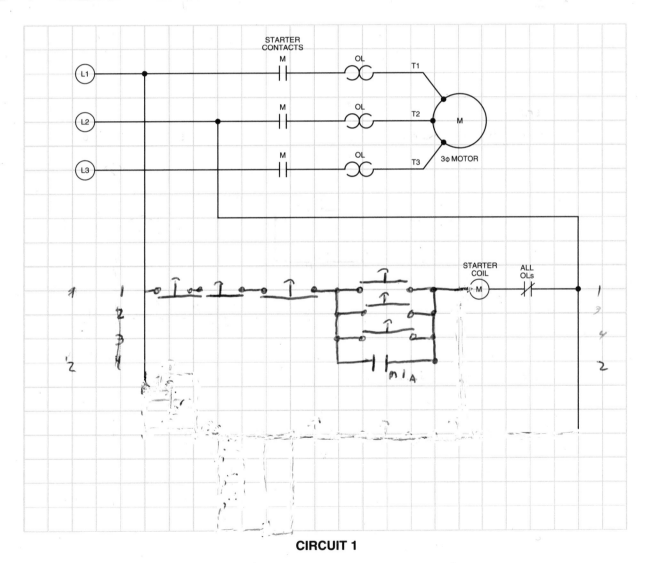

CIRCUIT 1

2. Redraw the control circuit line diagram of Circuit 1, adding red and green pilot lights. The red pilot light turns ON any time the motor is ON. The green pilot light turns ON any time the motor is OFF.

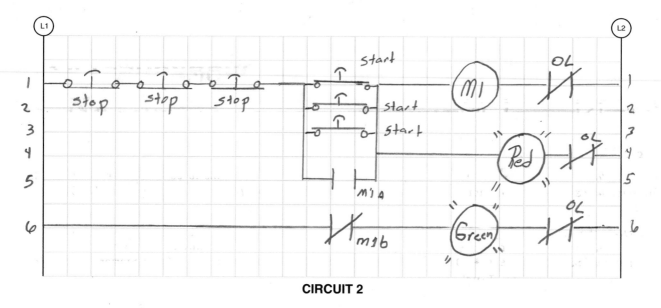

CIRCUIT 2

3. Redraw Circuit 2, adding a selector switch that is used to place the circuit in a jog or run position.

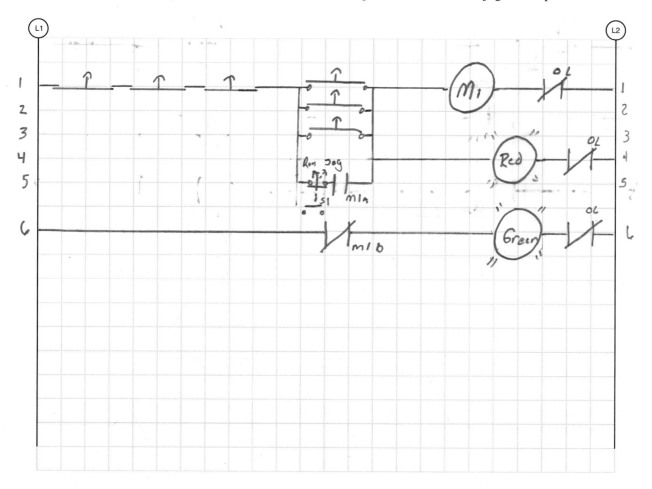

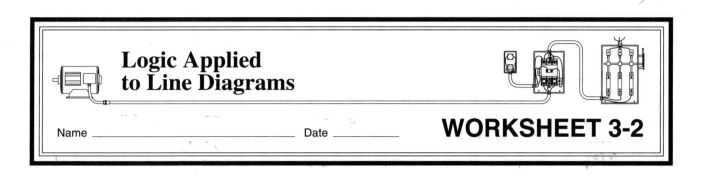

**Logic Applied
to Line Diagrams**

Name _____ Date _____

WORKSHEET 3-2

Circuit Overload Protection

Use standard lettering, line-reference, and cross-reference numbering systems to complete each line diagram. Wire-reference numbers are not required. Note: Assume that the motor starters have several NO and NC auxiliary contacts.

1. Draw the line diagram of three magnetic motor starters controlled by a common start/stop pushbutton station. Interconnect the three motor starters so if an overload occurs on any of the starters, all three are automatically disconnected. Design the circuit so Motor Starter 1 energizes Motor Starter 2, and Motor Starter 2 energizes Motor Starter 3.

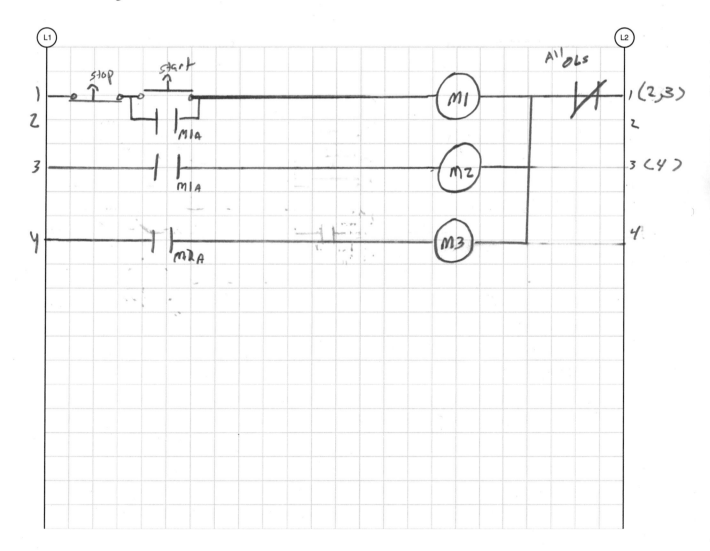

2. Draw the line diagram of three magnetic motor starters controlled by three individual start/stop pushbutton stations. Include a Master Stop Pushbutton that stops all three starters when pressed. Design the circuit so the starters can be individually stopped by each start/stop pushbutton station when the Master Stop Pushbutton is not used. Each starter must have its own overload protection.

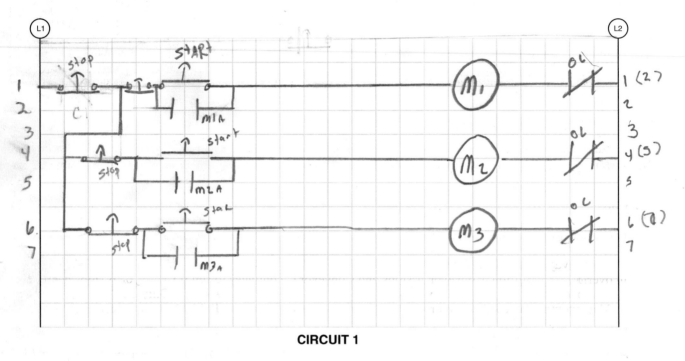

CIRCUIT 1

3. Redraw Circuit 1, adding a pressure switch that automatically stops all motors if excessive pressure is reached. Add a red pilot light that turns ON when Motor 1 is running, a green pilot light that turns ON when Motor 2 is running, and an amber pilot light that turns ON when Motor 3 is running.

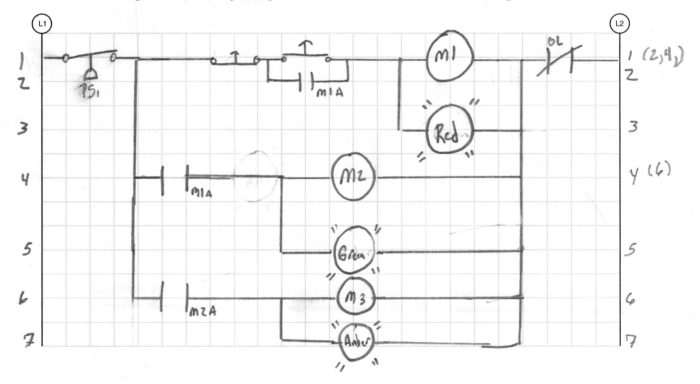

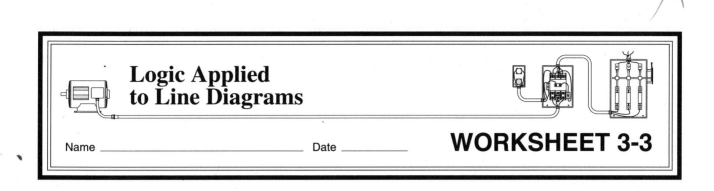

Multiple Conveyor Control

Use standard lettering, line-reference, and cross-reference numbering systems to complete each line diagram. Wire-reference numbers are not required. Note: *The motor starters include both NO and NC contacts in any number required to solve the problem.*

1. Complete the line diagram of a control circuit for a three-belt, three-motor conveyor system in which Conveyor A feeds bulk material to Conveyor B, Conveyor B feeds the material to Conveyor C, and Conveyor C dumps the material. To prevent material pileups and ensure safe operation, design the circuit so Conveyor A and Conveyor B cannot start unless Conveyor C starter is energized, Conveyor A cannot start unless Conveyor B starter is energized, and Conveyor A and Conveyor B stop if Conveyor C stops because of an overload. In addition, Conveyor A stops and Conveyor C continues to run if Conveyor B stops because of an overload. Conveyor B and Conveyor C continue to run if Conveyor A stops. Only one start and one stop pushbutton may be used to control the conveyor system. Include individual pilot lights to show which conveyors are running.

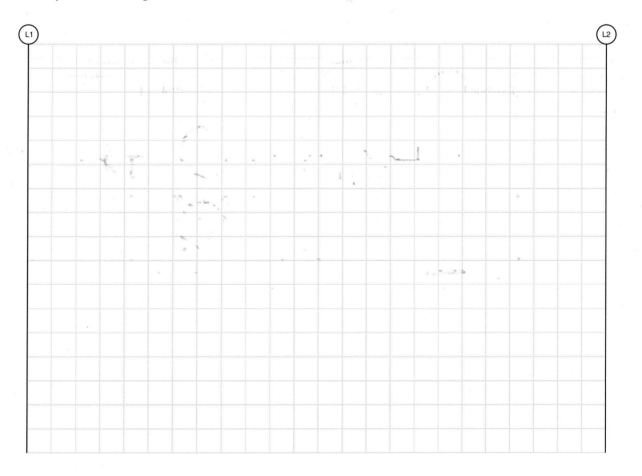

Logic Applied to Line Diagrams

Name _____ Date _____

WORKSHEET 3-4

Selector Switch Control

Use standard lettering, line-reference, and cross-reference numbering systems to complete each line diagram. Wire-reference numbers are not required. Note: *The motor starters include both NO and NC contacts in any number required to solve the problems.*

1. Design a start/stop/jog control circuit using a selector switch to provide the jog/run function. In the jog position, the start pushbuttons control the jogging of the motor. In the run position, the circuit functions as a standard start/stop circuit with memory. Include a red pilot light that turns ON when the circuit is in the jog condition and a green pilot light that turns ON when the circuit is in the run condition.

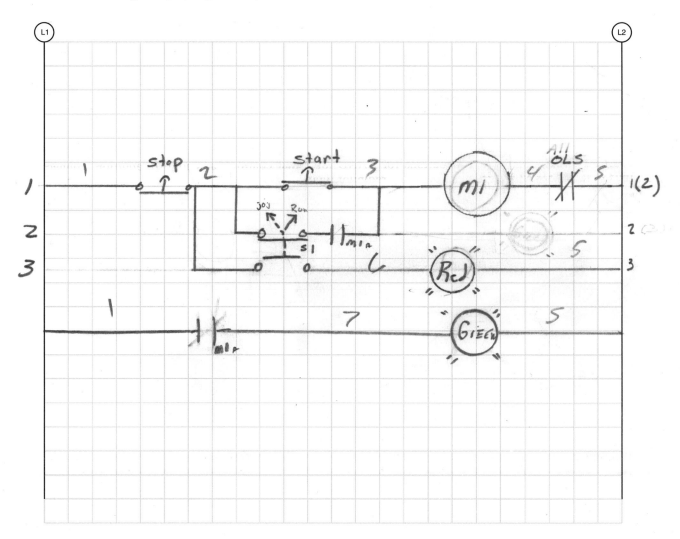

2. Design a motor control circuit that includes a standard start/stop pushbutton station that controls a motor starter. Include a green push-to-test pilot light that is used to indicate when a starter is energized and when the testing of the bulb is enabled by simply pushing the color cap on the pushbutton. The starter must not energize when testing the bulb.

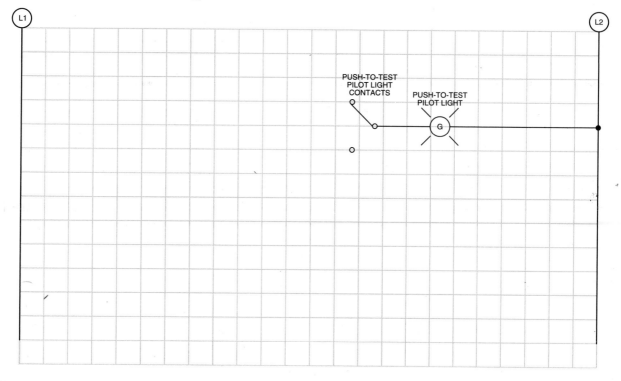

3. Design a circuit using two pushbuttons for NAND logic that control a solenoid. In addition, connect two limit switches for NOR logic that control a red pilot light.

Logic Applied to Line Diagrams

Name _____ Date _____

WORKSHEET 3-5

OR Circuit Logic

Use standard lettering, line-reference, and cross-reference numbering systems to complete each line diagram. Wire-reference numbers are not required.

1. Design an OR logic circuit in which one mechanical switch (limit), one manual switch (pushbutton), and two liquid level switches make up the signal section. The decision is OR logic and the action is a bell ringing and a red light ON simultaneously.

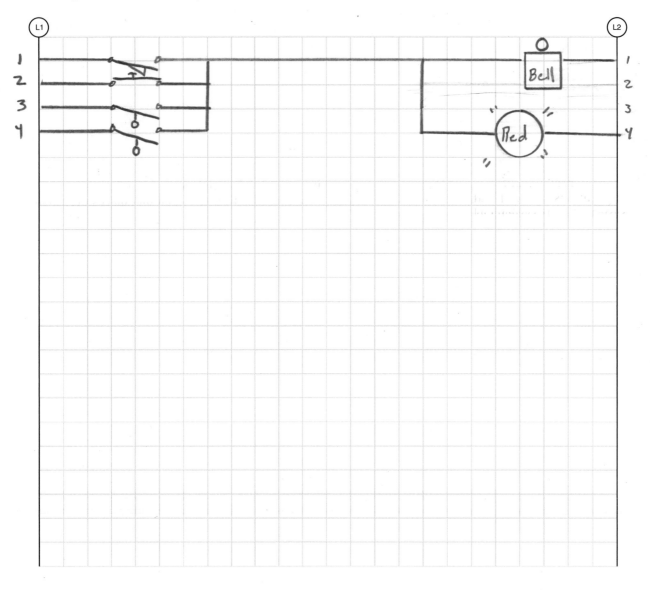

2. Develop a NOT logic circuit in which an automatic temperature control (temperature switch) is the signal, the decision is NOT logic, and the action is a red pilot light and a low-power heating element activated simultaneously.

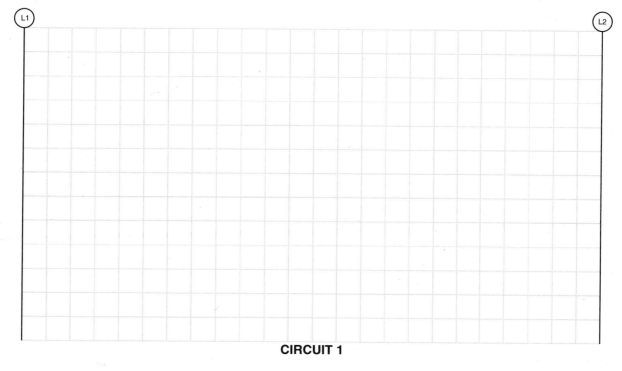

CIRCUIT 1

3. Redraw Circuit 1, adding a second temperature switch so either switch activates the loads when the temperature is below the set level. In addition, include a 2-position selector switch that places the circuit in a system OFF or system ON position.

Logic Applied to Line Diagrams

Name _____ Date _____

WORKSHEET 3-6

AND/OR Combination Logic

Use standard lettering, line-reference, and cross-reference numbering systems to complete each line diagram. Wire-reference numbers are not required.

1. Design a circuit with AND/OR combination logic so the signal is manual (two pushbuttons and two limit switches), the decision is AND/OR combination logic so at least three devices must be actuated, and the action is a siren and a red light activated simultaneously.

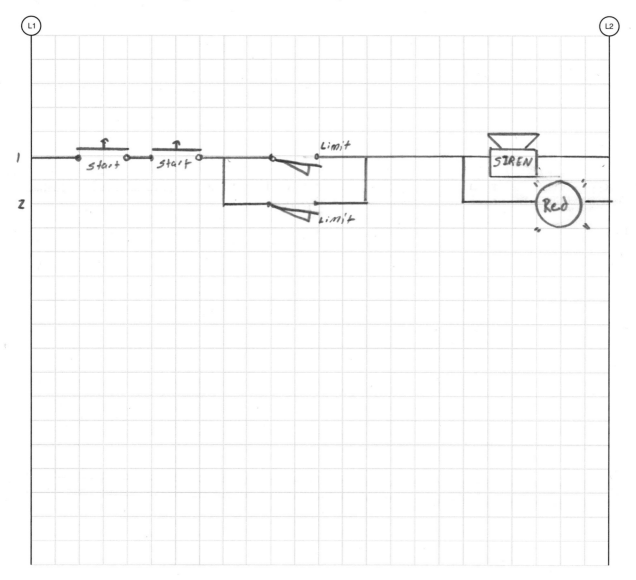

2. Design a circuit so the signal is automatic (vacuum switch) and manual (pushbutton), the decision is memory, so a pushbutton starts the operation and holds until a vacuum switch stops the operation, and the action is a magnetic starter coil and a red pilot light that indicates when the motor is energized.

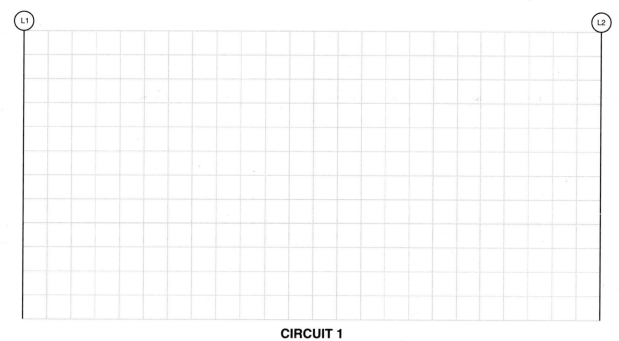

CIRCUIT 1

3. Redraw Circuit 1, adding a red pilot light that indicates when the motor is energized and a second pushbutton that stops the motor any time it is running.

Logic Applied to Line Diagrams

Name _____ Date _____

WORKSHEET 3-7

Circuit Logic

Match each statement to the proper circuit. All circuits have been drawn with a light (L) to represent the load, whether it is a motor, bell, light, or any other load. In addition, each switch is illustrated as a pushbutton whether it is a maintained switch, momentary contact switch, pushbutton, switch-ON target, or any other type of switch.

_____C_____ 1. The warning light outside a darkroom (DO NOT ENTER) is OFF when the white light in the darkroom is ON. The warning light outside is ON when the white light in the darkroom is OFF.

_____I_____ 2. Switches are connected so the canopy of an airplane is ejected first and the pilot second regardless of which switch the pilot activates first.

_____E_____ 3. Two guns are connected to individual targets by switches so when two people compete in firing them, the fastest to fire is shown.

_____A_____ 4. Switches are connected so several people are required to fire a missile.

_____B_____ 5. An indicating light is ON to warn of danger when an oven is ON.

_____D_____ 6. A security guard monitoring a light panel can tell if the front door, back door, or both doors are open.

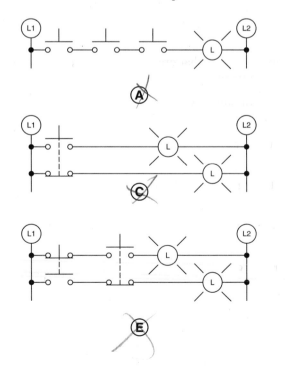

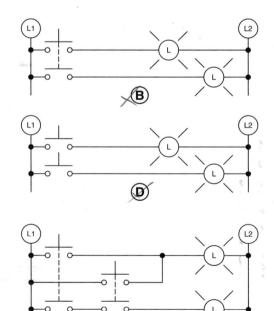

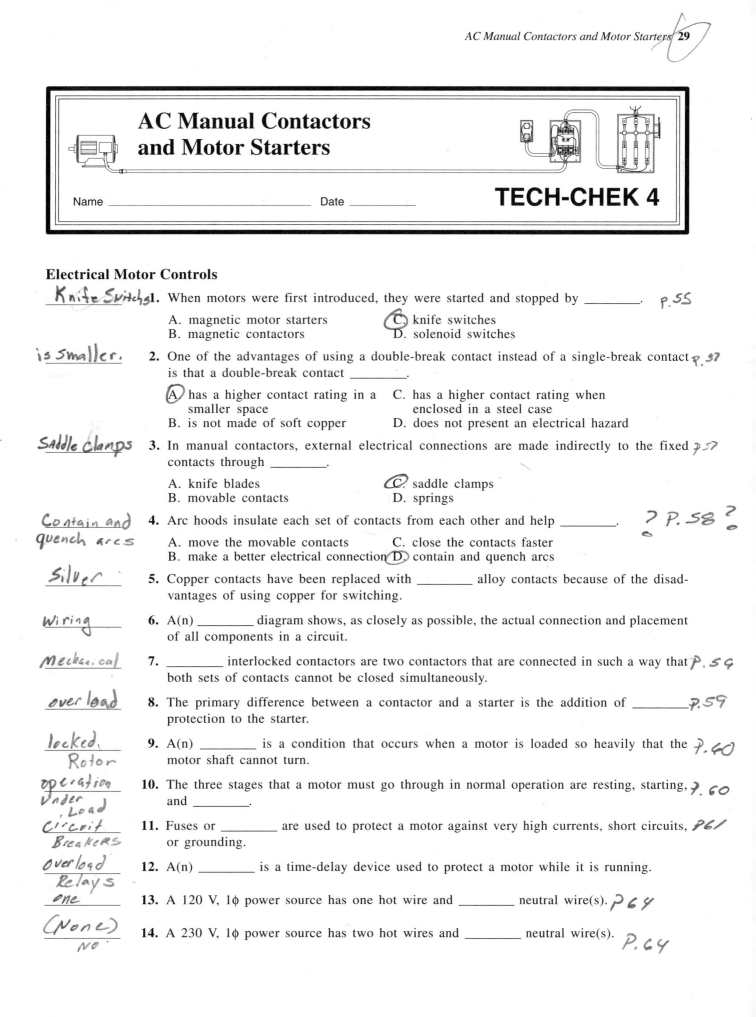

AC Manual Contactors and Motor Starters

Name _____ Date _____

TECH-CHEK 4

Electrical Motor Controls

Knife Switches 1. When motors were first introduced, they were started and stopped by _____. *p.55*

    A. magnetic motor starters   C. knife switches
    B. magnetic contactors   D. solenoid switches

is smaller. 2. One of the advantages of using a double-break contact instead of a single-break contact *p.57* is that a double-break contact _____.

    A. has a higher contact rating in a   C. has a higher contact rating when
        smaller space   enclosed in a steel case
    B. is not made of soft copper   D. does not present an electrical hazard

Saddle Clamps 3. In manual contactors, external electrical connections are made indirectly to the fixed *p.57* contacts through _____.

    A. knife blades   C. saddle clamps
    B. movable contacts   D. springs

Contain and quench arcs 4. Arc hoods insulate each set of contacts from each other and help _____. *? P.58 ?*

    A. move the movable contacts   C. close the contacts faster
    B. make a better electrical connection   D. contain and quench arcs

Silver 5. Copper contacts have been replaced with _____ alloy contacts because of the disadvantages of using copper for switching.

Wiring 6. A(n) _____ diagram shows, as closely as possible, the actual connection and placement of all components in a circuit.

Mechanical 7. _____ interlocked contactors are two contactors that are connected in such a way that *P.59* both sets of contacts cannot be closed simultaneously.

overload 8. The primary difference between a contactor and a starter is the addition of _____ *P.59* protection to the starter.

locked Rotor 9. A(n) _____ is a condition that occurs when a motor is loaded so heavily that the *P.60* motor shaft cannot turn.

operation under Load 10. The three stages that a motor must go through in normal operation are resting, starting, *P.60* and _____.

Circuit Breakers 11. Fuses or _____ are used to protect a motor against very high currents, short circuits, *P61* or grounding.

overload Relays 12. A(n) _____ is a time-delay device used to protect a motor while it is running.

one 13. A 120 V, 1φ power source has one hot wire and _____ neutral wire(s). *P64*

(None) No 14. A 230 V, 1φ power source has two hot wires and _____ neutral wire(s). *P.64*

three   **15.** A 3ϕ power source has _____ hot wire(s) and zero neutral wires. p64

disconnect   **16.** A(n) _____ is a device that is used only periodically to remove electrical circuits from their supply source. p56

lighting   **17.** Manual contactors are normally used with _____ circuits and resistance loads. p 56

Heat coil   **18.** The main device in an overload relay is the _____.

Contactor   **19.** A manual starter is a(n) _____ with an added overload protection device. p60

NEC®   **20.** Overload protection devices are required by the _____.

Ambient   **21.** _____ temperature is the temperature of the air surrounding a motor. p 60

eutectic   **22.** A(n) _____ alloy is a metal which melts at a fixed temperature p-61

overload   **23.** A starter may be reset after a(n) _____ is removed. p.61

Mechanical **24.** Enclosures provide _____ and electrical protection for the operator and the starter. p.65

Manual
Contactors   **25.** A(n) _____ is a control device that uses pushbuttons to energize or de-energize the load connected to it. P.56

Double-Break **26.** _____ contacts are contacts that break the electrical circuit in two places. p57

heater-coil   **27.** A(n) _____ is a sensing device used to monitor the heat generated by excessive current and the heat created through ambient temperature rise. p61

one   **28.** NEMA Type __1__ enclosures are intended for indoor use primarily to provide a degree of protection against human contact with the enclosed equipment in locations where unusual service conditions do not exist. P 65

dRill presses **29.** Manual motor starters are used in applications such as air compressors, conveyor systems, and _____. p65

(_electrical_)
(_motor_)   **30.** Manual contactors directly control power _____ circuits. p.59

AC Manual Contactors and Motor Starters

Name _____ Date _____

WORKSHEET 4-1

NEMA Enclosures

Determine the required NEMA enclosure for each location.

_____ **1.** Required enclosure is NEMA _____.

_____ **2.** Required enclosure is NEMA _____.

_____ **3.** Required enclosure is NEMA _____.

_____ **4.** Required enclosure is NEMA _____.

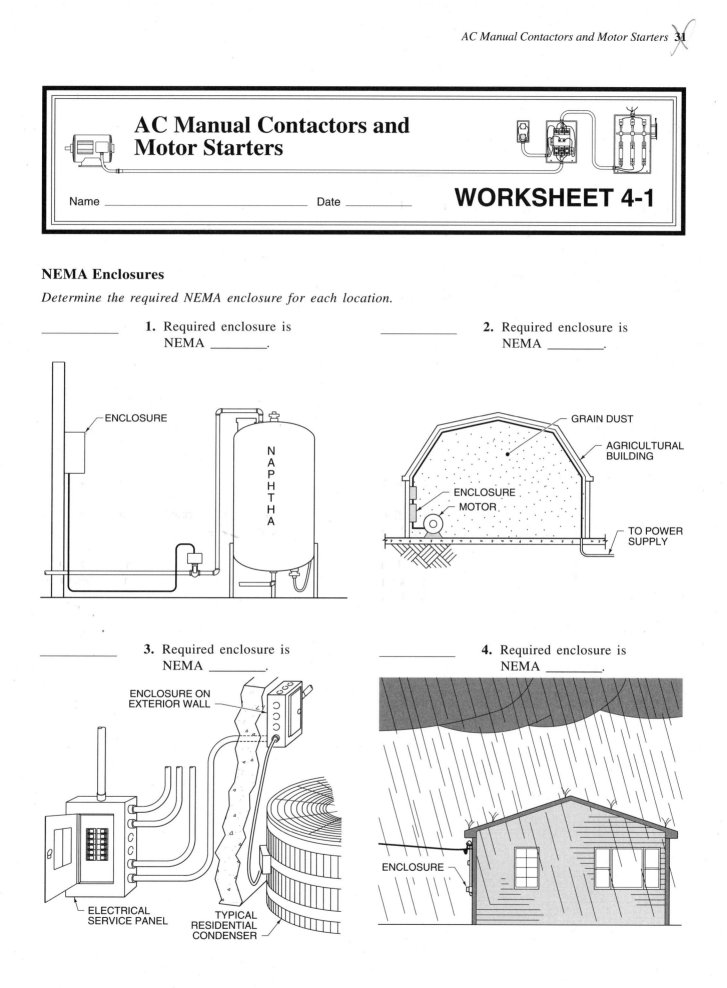

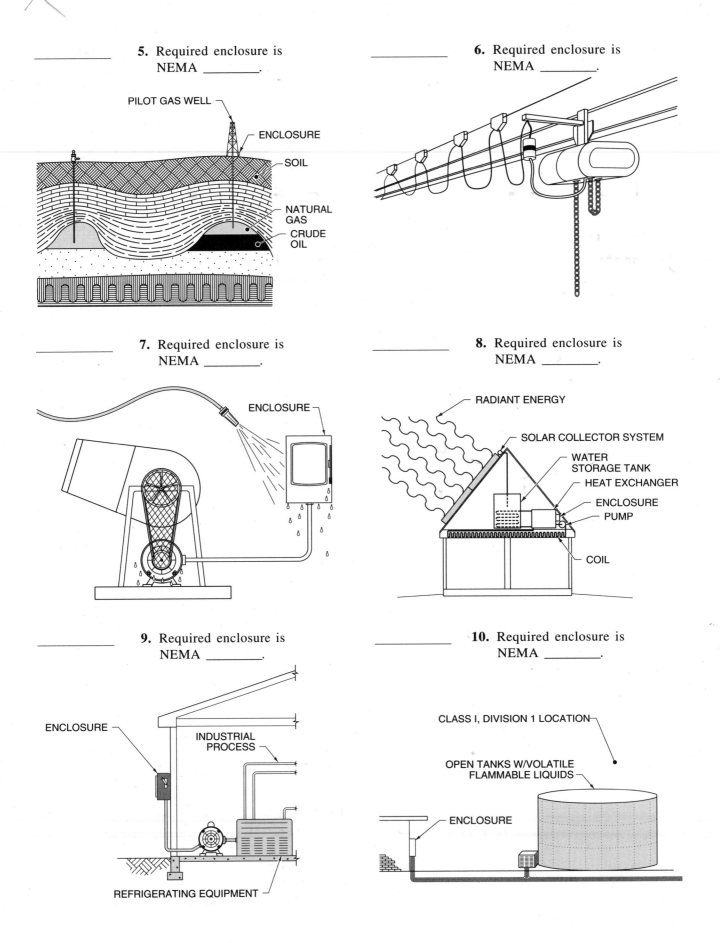

_____ **5.** Required enclosure is NEMA _____.

PILOT GAS WELL
ENCLOSURE
SOIL
NATURAL GAS
CRUDE OIL

_____ **6.** Required enclosure is NEMA _____.

_____ **7.** Required enclosure is NEMA _____.

ENCLOSURE

_____ **8.** Required enclosure is NEMA _____.

RADIANT ENERGY
SOLAR COLLECTOR SYSTEM
WATER STORAGE TANK
HEAT EXCHANGER
ENCLOSURE
PUMP
COIL

_____ **9.** Required enclosure is NEMA _____.

ENCLOSURE
INDUSTRIAL PROCESS
REFRIGERATING EQUIPMENT

_____ **10.** Required enclosure is NEMA _____.

CLASS I, DIVISION 1 LOCATION
OPEN TANKS W/VOLATILE FLAMMABLE LIQUIDS
ENCLOSURE

AC Manual Contactors and Motor Starters

Name _____ Date _____

WORKSHEET 4-2

Switching Arrangements

Use Data Sheet A to complete the 10 basic switching arrangements that are possible with a dual-element coil and dual-voltage power supply. Draw a schematic diagram of each circuit to the right of each switching arrangement.

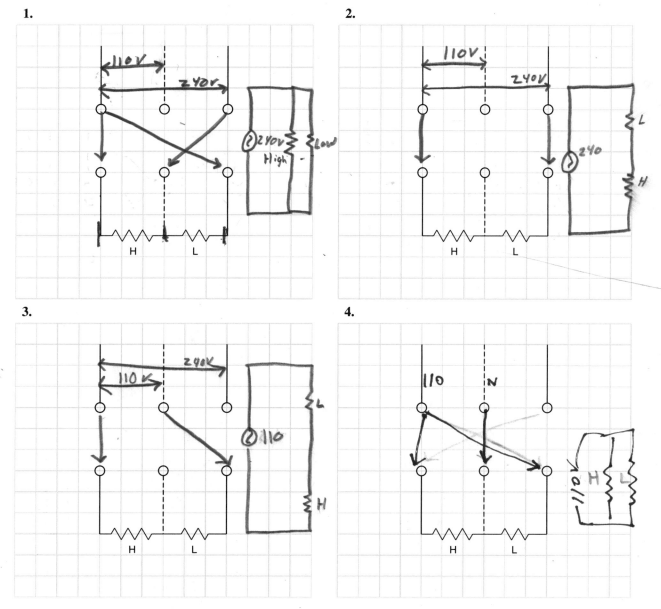

1.

2.

3.

4.

5.

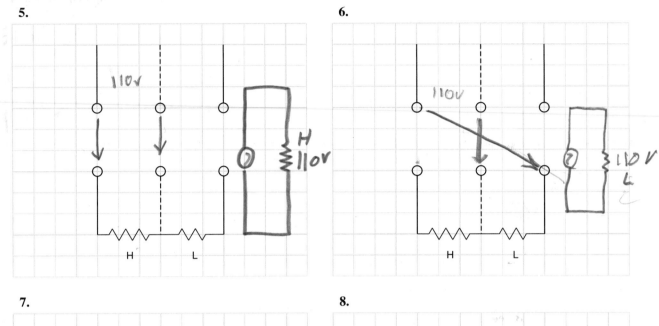

6.

7.

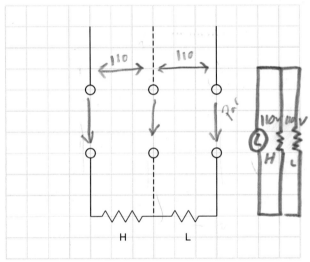

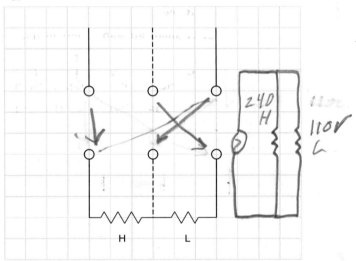

8.

9.

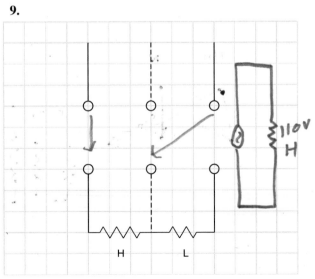

10.

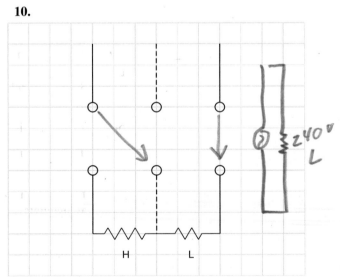

**Magnetism and
Magnetic Solenoids**

Name _____ Date _____

TECH-CHEK 5

Electrical Motor Controls

(D)

__A, B, & C__ **1.** The strength of an electromagnet can be increased by _____.

    A. increasing the voltage   C. inserting an iron core
    B. increasing the number of   D. A, B, and C
       turns of wire

__B__ **2.** The solenoid armature is made from thin laminated pieces to help reduce _____.

    A. air gaps   C. slow action
    B. eddy currents   D. cost

__B__ **3.** A small air gap is left in the iron core armature circuit to _____.

    A. prevent eddy currents   C. reduce cost
    B. prevent the armature from   D. prevent heat buildup in the armature
       staying in a sealed position   circuit

__B__ **4.** A shading coil is added to the armature core to _____.

    A. reduce air gaps   C. prevent eddy currents
    B. keep the armature   D. prevent the armature from staying
       firmly seated   in a sealed position

? __A__ **5.** Excessively noisy solenoids may be a result of _____.

    A. a broken shading coil   C. dirt, rust, or filings on the magnetic face
    B. voltage too low   D. A, B, and C

__Permanent__ **6.** _____ magnets are magnets that can retain their magnetism after a magnetizing force has been removed.

__Temporary__ **7.** _____ magnets are magnets that have extreme difficulty in retaining any magnetism after the magnetizing force has been removed.

__Inrush__ **8.** _____ current is the current that a solenoid coil draws when first turned ON.

__Sealed__ **9.** _____ current is the current that a solenoid coil draws after the armature circuit is closed.

__Volt-Amps__ **10.** The amount of force a solenoid can deliver is normally rated in _____.

__Duty Cycle__ **11.** _____ is the number of times a solenoid can operate in a given time (normally per minute).

__Molecular__ **12.** The _____ theory of magnetism states that all substances are made up of an infinite number of molecular magnets.

__Solenoid__ **13.** A(n) _____ is an electromagnet consisting of a coil of wire and a source of voltage.

__Plunger__ **14.** A(n) _____ solenoid contains only a moving iron cylinder.

Eddy Current **15.** _____ is unwanted current induced in the metal structure of a device.

Shading Coil **16.** A(n) _____ is a single turn of conducting material mounted on the face of the magnetic laminate assembly or armature.

epoxy resin **17.** The mechanical life of most coils is improved by encapsulating them in _____ or a glass-reinforced alkyd material.

Pickup **18.** _____ voltage is the minimum control voltage which causes the armature to start to move.

Drop-out **19.** _____ voltage is the voltage which exists when the voltage has reduced sufficiently to allow the solenoid to open.

Number **20.** Manufacturers provide letter or _____ codes to indicate the voltages available for a given solenoid.

Direct-acting **21.** _____ 2-way valves are common in refrigeration equipment.

Low Voltage **22.** _____ protection is accomplished by a continuous-duty solenoid that is energized whenever the line voltage is present.

Overheats **23.** A solenoid _____ when the voltage is excessive.

Excessive **24.** _____ voltages may damage the insulation on the solenoid coil.

ohmeter **25.** No movement of the needle on an analog meter or infinite resistance on a(n) _____ meter indicates the coil is open and defective.

Solenoid Identification

_____ C   **1.** Vertical-action

_____ D   **2.** Horizontal-action

_____ A   **3.** Clapper

_____ B   **4.** Bell-crank

_____ E   **5.** Plunger

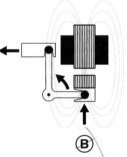

(A)

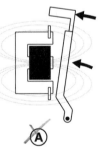

(B)

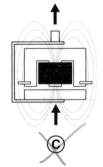

(C)

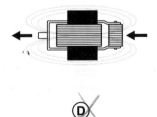

(D)

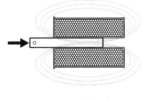

(E)

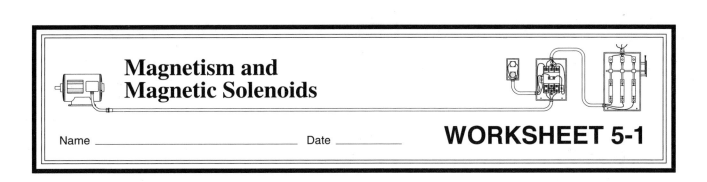

Magnetism and Magnetic Solenoids

Name _____ Date _____

WORKSHEET 5-1

Pushbutton Circuit Control

Complete the line diagram of the cylinder control circuit. Use standard lettering, numbering, and coding information.

1. Draw the line diagram of the control circuit using one pushbutton (PB1) to control the advance of the cylinder and another pushbutton (PB2) to control the retracting of the cylinder.

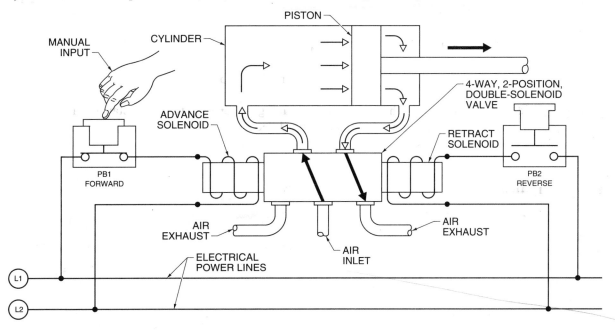

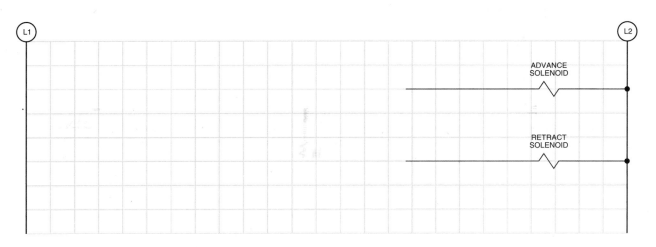

Magnetism and Magnetic Solenoids

Name _____ Date _____

WORKSHEET 5-2

Manual Control Circuit with Memory

Complete the line diagram of the cylinder control circuit. Use standard lettering, numbering, and coding information.

1. Complete the line diagram so the cylinder advances when PB1 is pressed and released. Include a limit switch that automatically returns the cylinder when the cylinder is fully advanced. *Note:* A contactor which is capable of controlling NO and NC contacts is provided in parallel with the solenoid. The contactor can be used to add circuit memory.

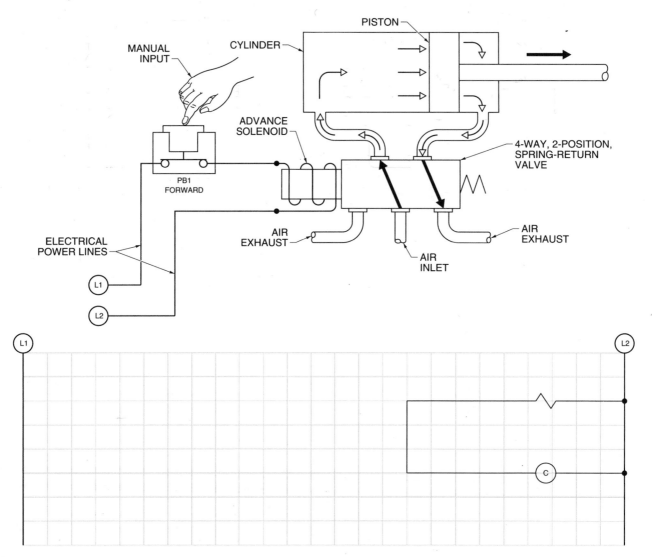

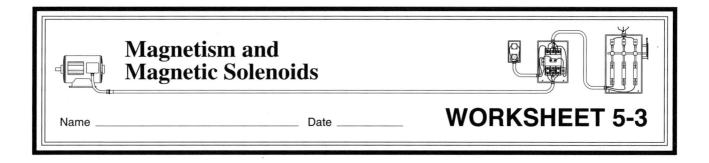

Magnetism and Magnetic Solenoids

Name _____ Date _____

WORKSHEET 5-3

Automatic Circuit Control

Complete the line diagram of the cylinder control circuit. Use standard lettering, numbering, and coding information.

1. Complete the line diagram so PB1 and a foot switch advance the cylinder and PB2 or a limit switch retracts the cylinder.

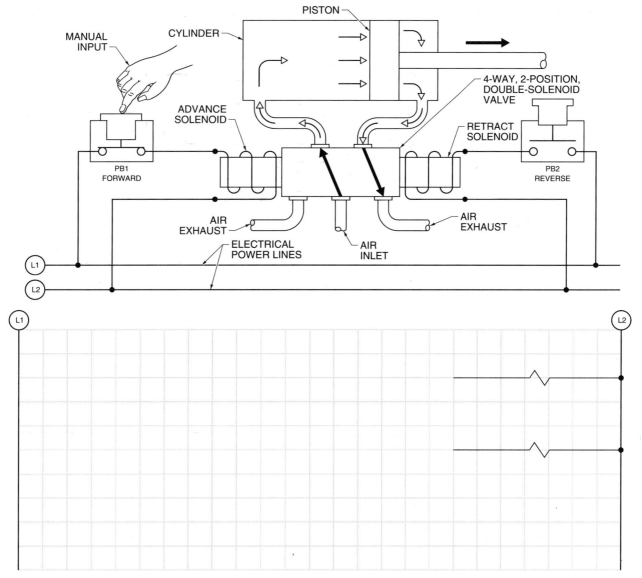

Magnetism and Magnetic Solenoids

Name _____ Date _____

WORKSHEET 5-4

Manual Override Circuit Control

Complete the line diagram of the cylinder control circuit. Use standard lettering, numbering, and coding information.

1. Design the circuit so the cylinder advances only if two pushbuttons (PB1 and PB2) are pressed. Add a contactor that automatically returns the cylinder. Include a manual return pushbutton (PB3) to manually return the cylinder if the pressure falls below the setting of the pressure switch.

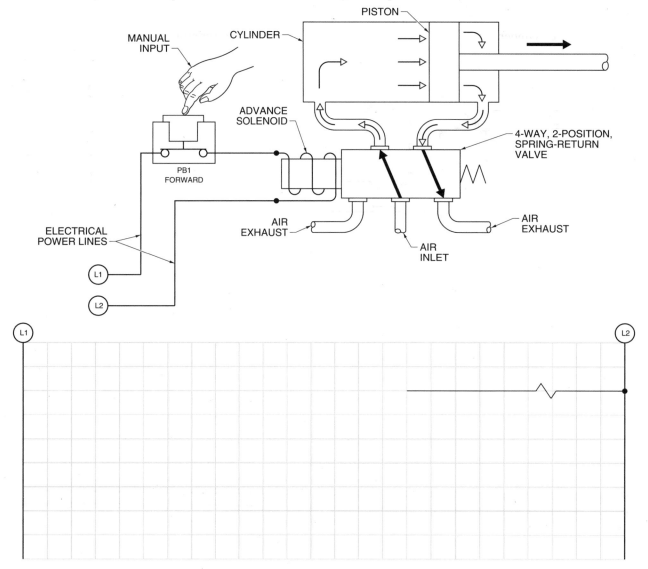

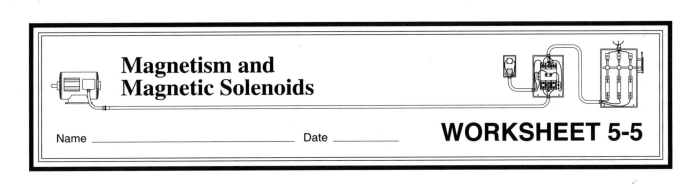

Magnetism and Magnetic Solenoids

Name _____ Date _____

WORKSHEET 5-5

Selector Switch Circuit Control

Draw the line diagram of the cylinder control circuit. Use standard lettering, numbering, and coding information.

1. Add a selector switch to determine manual or automatic control of the cylinder. The cylinder cycles continuously and automatically back and forth when the selector switch is in the automatic position and the start pushbutton is pressed and released. The cycling stops when the stop pushbutton is pressed. The cylinder advances only when a pushbutton (PB1) is pressed and retracts only when a second pushbutton (PB2) is pressed when the selector switch is in the manual position. *Note:* Limit switch 1 (LS1) is NO (held closed) because the cylinder is retracted. Assume that no memory is required for either solenoid.

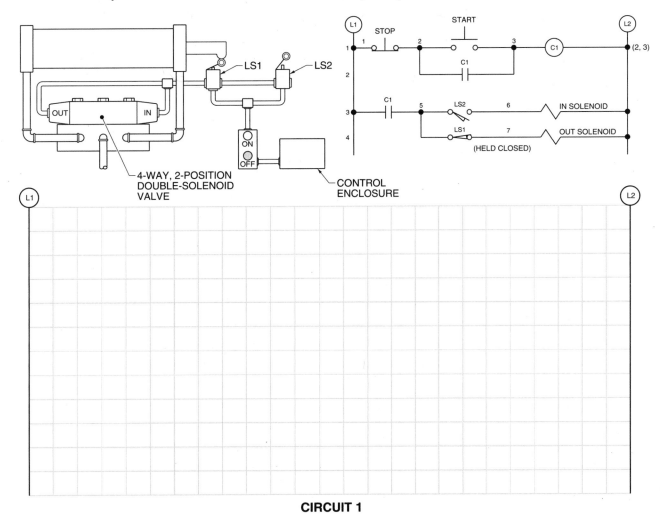

CIRCUIT 1

2. Redraw Circuit 1, adding an emergency stop pushbutton that stops the solenoids from being energized in either the manual or automatic condition until a reset pushbutton is activated. The circuit includes six pushbuttons (start, stop, PB1, PB2, reset, and emergency stop), two limit switches, two solenoids, a selector switch, and two contactors.

CIRCUIT 2

3. Redraw Circuit 2, adding a red light to indicate when the in solenoid is energized, a green light to indicate when the out solenoid is energized, and a yellow light to indicate when the emergency stop is pressed.

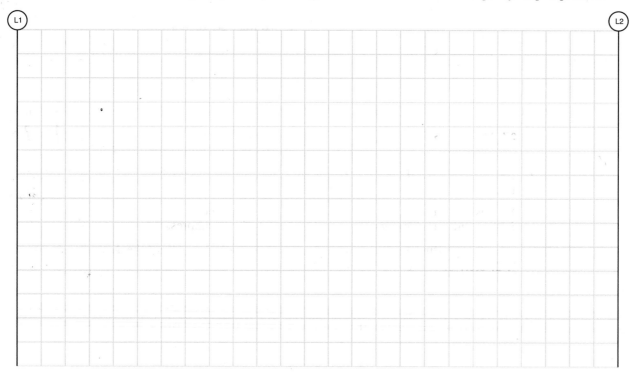

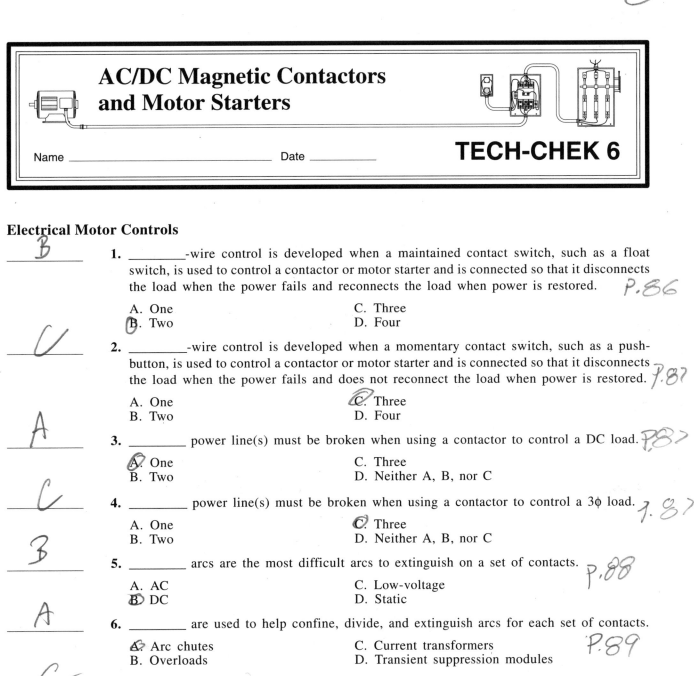

AC/DC Magnetic Contactors and Motor Starters

Name _____ Date _____

TECH-CHEK 6

Electrical Motor Controls

B 1. _____-wire control is developed when a maintained contact switch, such as a float switch, is used to control a contactor or motor starter and is connected so that it disconnects the load when the power fails and reconnects the load when power is restored. *P.86*

    A. One   C. Three
    B. Two   D. Four

C 2. _____-wire control is developed when a momentary contact switch, such as a push-button, is used to control a contactor or motor starter and is connected so that it disconnects the load when the power fails and does not reconnect the load when power is restored. *P.87*

    A. One   C. Three
    B. Two   D. Four

A 3. _____ power line(s) must be broken when using a contactor to control a DC load. *P.87*

    A. One   C. Three
    B. Two   D. Neither A, B, nor C

C 4. _____ power line(s) must be broken when using a contactor to control a 3ϕ load. *P.87*

    A. One   C. Three
    B. Two   D. Neither A, B, nor C

B 5. _____ arcs are the most difficult arcs to extinguish on a set of contacts. *P.88*

    A. AC   C. Low-voltage
    B. DC   D. Static

A 6. _____ are used to help confine, divide, and extinguish arcs for each set of contacts. *P.89*

    A. Arc chutes   C. Current transformers
    B. Overloads   D. Transient suppression modules

C 7. _____ are used to provide a magnetic field that helps move contacts apart as quickly as possible. *P.90*

    A. Silver contacts   C. Blow-out coils
    B. Current transformers   D. Transient suppression modules

A 8. The power rating of a contactor or motor starter _____ as the NEMA number (Size 1, 2, etc.) of the contactor or motor starter increases. *P.91*

    A. increases   C. remains the same
    B. decreases   D. Neither A, B, nor C

A 9. The current rating of a contactor or motor starter is the rating for _____. *P.91*

    A. each individual contact   C. 12 hr of operation
    B. the whole contactor divided by   D. the whole contactor
        the number of contacts

overloads   **10.** The main difference between a contactor and a motor starter is the addition of _____ to the motor starter. P.92

Magnetic   **11.** Two overload relays used to protect motors are thermal and _____ overload relays.

Service Factor   **12.** Ambient temperature, full-load current rating, and _____ must be known when selecting the overloads for a motor starter. P.95

12.5 Amps **13.** The amount of current a motor can safely draw for a short period of time is _____ A if the motor has a service factor of 1.25 and a current rating of 10 A. P.96

three   **14.** _____ individual overloads are required on a 3φ motor installation. p.98

Overload   **15.** An inherent motor protector is designed to protect a motor from _____. P.100

power poles   **16.** Fuses and _____ are two optional modifications that can be added to a contactor or motor starter. P.101

Coutactor   **17.** A(n) _____ is a control device that uses a small control current to energize or de-energize the load connected to it and has no overload protection.

AC   **18.** _____ contactor assemblies may have several sets of contacts. P.87

AC   **19.** _____ contactor assemblies are made of laminated steel. P.87
- A. AC
- B. DC
- C. Low-voltage
- D. Static

Arc Suppressor **20.** A(n) _____ is a device that dissipates the energy present across opening contacts. P.88

Bimetallic   **21.** A(n) _____ relay is an overload relay which resets automatically. P.94

Trip Indicator **22.** A(n) _____ is built into an overload device to indicate to the operator that an overload has taken place within the device. P.95

Current transformer **23.** _____ are used on large motor starters to reduce the current flowing to the overload relay.

less   **24.** As ambient temperature increases, _____ current is needed to trip overload devices. P.90
- A. more
- B. less
- C. does not change
- D. neither A, B, nor C

Inherent motor Protectors **25.** _____ are overload devices located directly on or in a motor to provide overload protection.

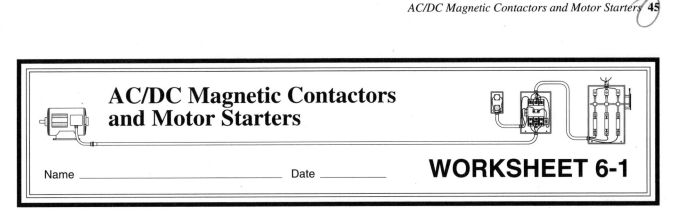

AC/DC Magnetic Contactors and Motor Starters

Name _____ Date _____

WORKSHEET 6-1

Wire Reference Numbers

Use Data Sheet B to assign a wire number to each wire in the line diagrams. Write each number directly above the wire.

1. The circuit is designed so that two magnetic motor starters are operated by two start/stop pushbutton stations with a common emergency stop. Assign the proper wire numbers to the wires of the two pilot lights so that when connected to the circuit, the red pilot light glows when magnetic motor starter M1 is ON and the green pilot light glows when magnetic motor starter M2 is ON.

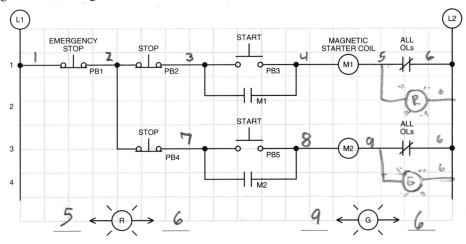

2. The circuit is designed with pushbuttons arranged for a sequence control of two motor starters. Assign the proper wire numbers to the wires of the foot switch and pushbutton so that when connected to the circuit, the foot switch starts coil M1 and the pushbutton starts coil M2.

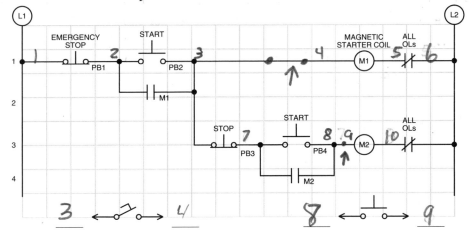

AC/DC Magnetic Contactors and Motor Starters

Name _____ Date _____

WORKSHEET 6-2

Basic Motor Control

Complete the wiring diagram according to the line diagram. Do not make any wire splices or additional terminal connections on the wiring diagram. All connections should run from terminal screw to terminal screw.

1. Draw the wiring diagram of the start/stop pushbutton station with memory.

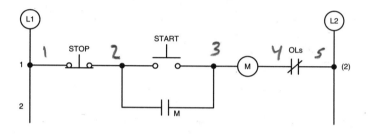

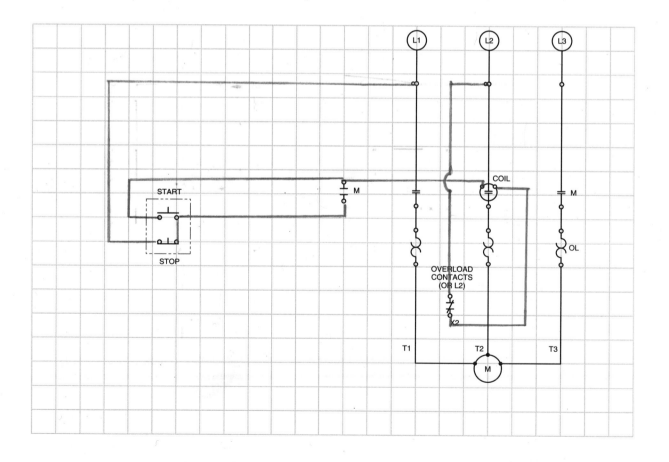

AC/DC Magnetic Contactors and Motor Starters

Name _____ Date _____

WORKSHEET 6-3

Motor Control with Pilot Light

Complete the wiring diagram according to the line diagram. Do not make any wire splices or additional terminal connections on the wiring diagram. All connections should run from terminal screw to terminal screw.

1. Draw the wiring diagram of the start/stop pushbutton station with memory and a pilot light that turns ON when the motor is not running. There is overload protection for the motor.

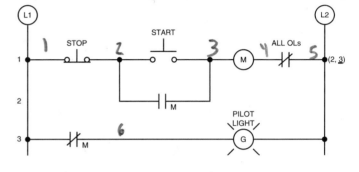

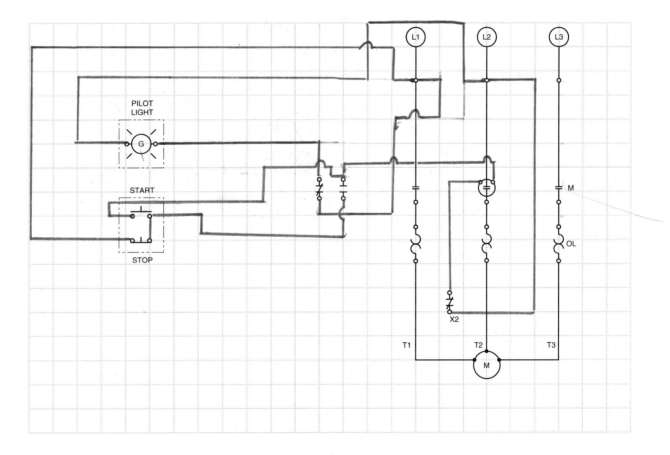

AC/DC Magnetic Contactors and Motor Starters

Name _____ Date _____

WORKSHEET 6-4

Multiple Control Stations

Complete the wiring diagram according to the line diagram. Do not make any wire splices or additional terminal connections on the wiring diagram. All connections should run from terminal screw to terminal screw.

1. Draw the wiring diagram of the three start/stop pushbutton stations with memory that control a single motor starter. There is overload protection for the motor.

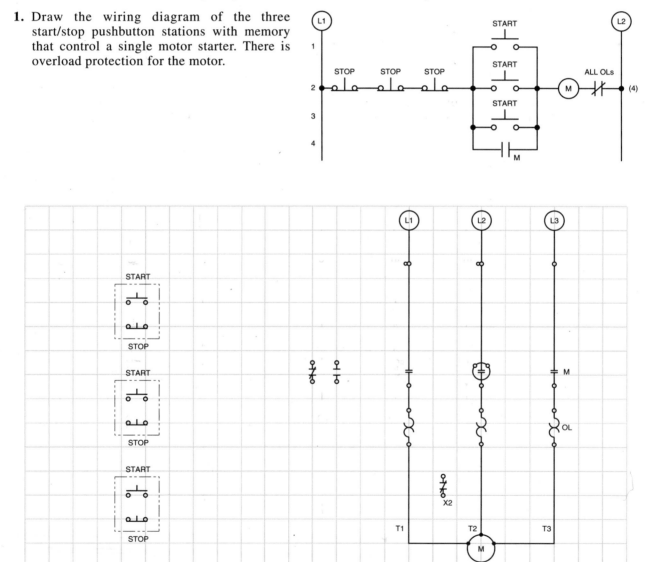

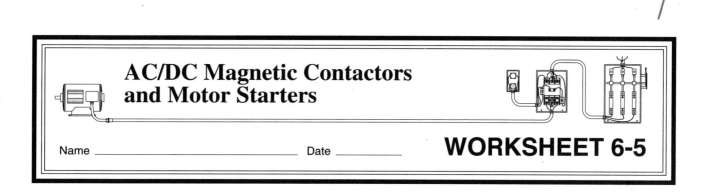

AC/DC Magnetic Contactors and Motor Starters

Name _____ Date _____

WORKSHEET 6-5

Multiple Motor Control

Complete the wiring diagram according to the line diagram. Do not make any wire splices or additional terminal connections on the wiring diagram. All connections should run from terminal screw to terminal screw.

1. Draw the wiring diagram of the start/stop pushbutton station with memory that controls two motor starters. The motor starters are wired so that if a maintained overload occurs on either one, both are automatically disconnected from the line. This is accomplished by wiring the holding circuit of each motor starter through the auxiliary contacts of the other. The circuit also provides for a sequence start of each motor to avoid the problems that could arise from both motors starting simultaneously.

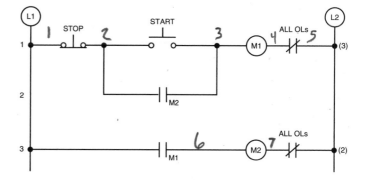

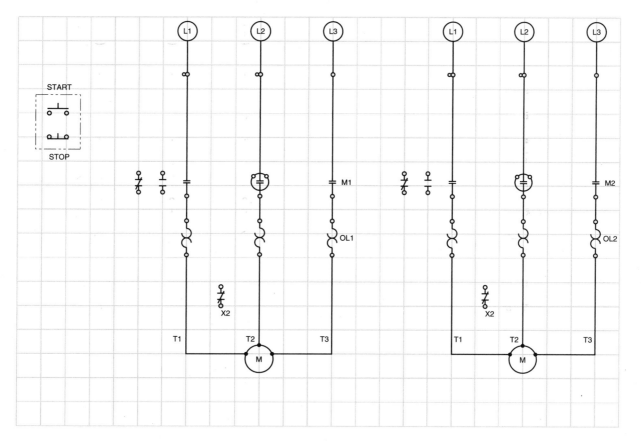

AC/DC Magnetic Contactors and Motor Starters

Name _____ Date _____

WORKSHEET 6-6

Multiple Control of Multiple Motors

Complete the wiring diagram according to the line diagram. Do not make any wire splices or additional terminal connections on the wiring diagram. All connections should run from terminal screw to terminal screw.

1. Draw the wiring diagram of the two separate start/stop pushbutton stations with memory that control two separate motor starters. A master stop pushbutton is included to turn OFF both motor starters. The overload relays on both motor starters are wired in series so that both drop out when a maintained overload occurs in either one.

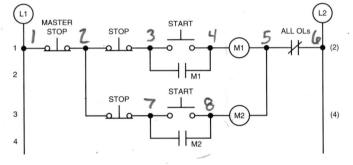

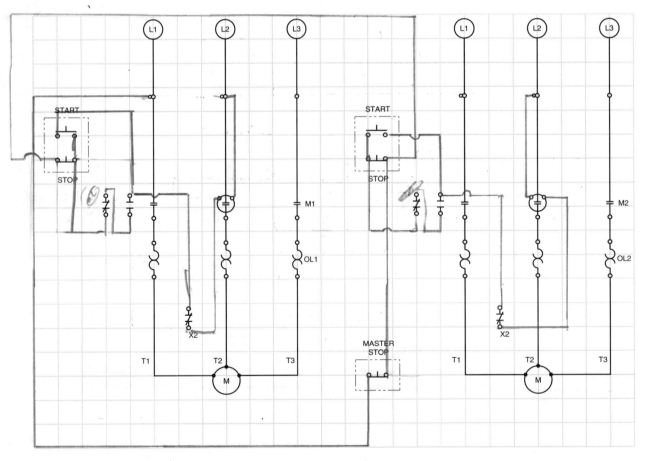

AC/DC Magnetic Contactors and Motor Starters

Name _____ Date _____

WORKSHEET 6-7

Circuit Wiring

Use Data Sheet C to wire the equipment. All wire splices should be inside the enclosures and conduit. Do not make any wire splices that are not necessary.

1. Wire the equipment as required by the conduit and enclosure arrangement. Power feed is through the start/stop enclosure.

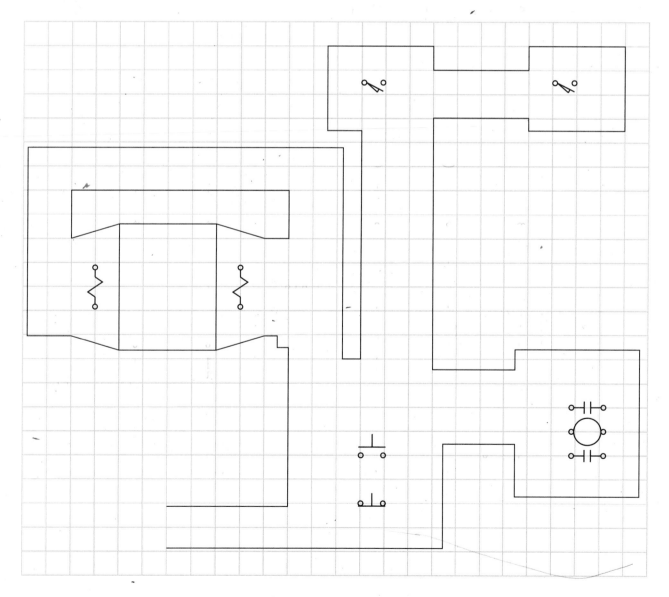

2. Wire the equipment as required by the conduit and enclosure arrangement. Power feed is through the limit switch enclosure.

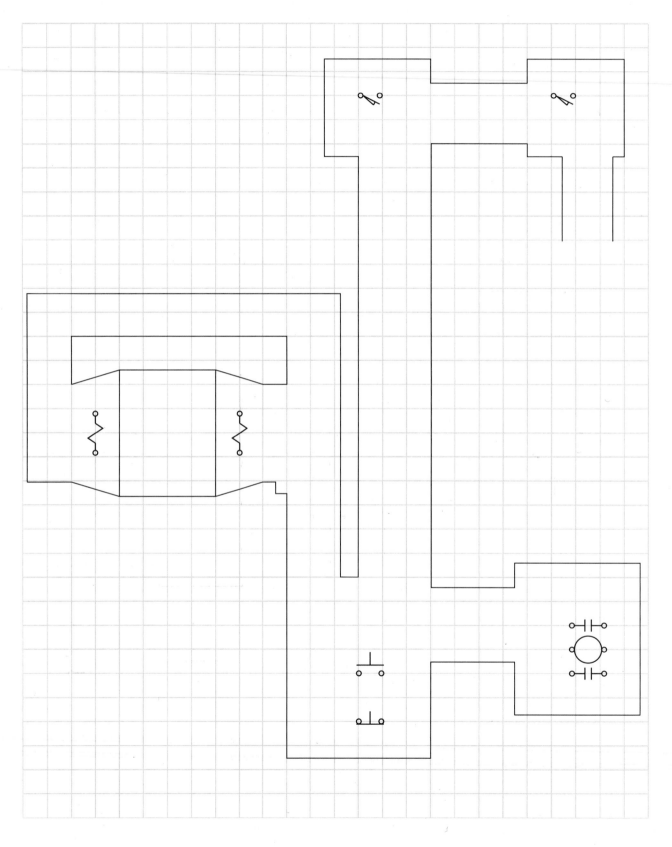

3. Wire the equipment as required by the conduit and enclosure arrangement. Power feed is through the control relay enclosure.

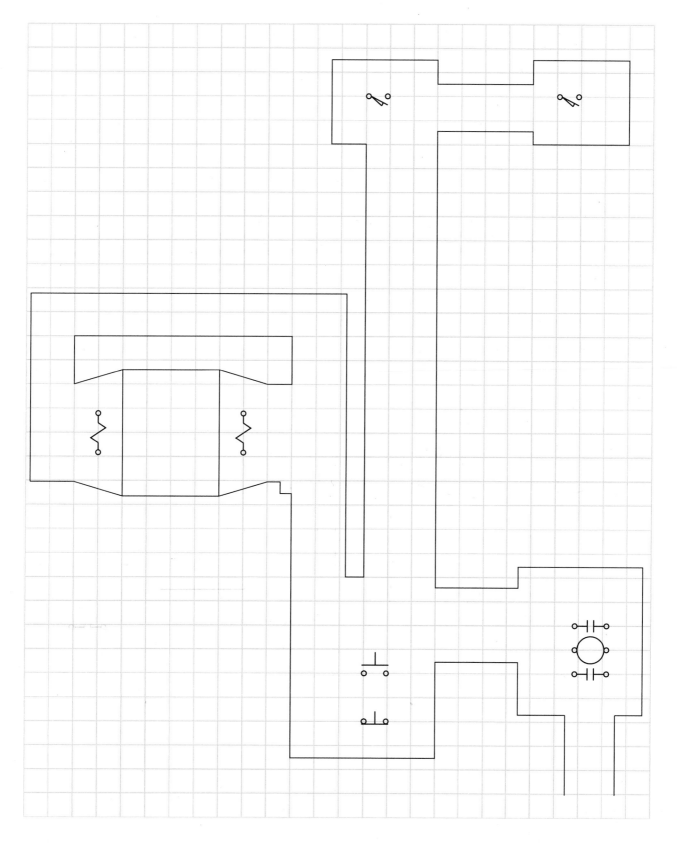

4. Wire the equipment as required by the conduit and enclosure arrangement. Power feed is through the solenoid valve.

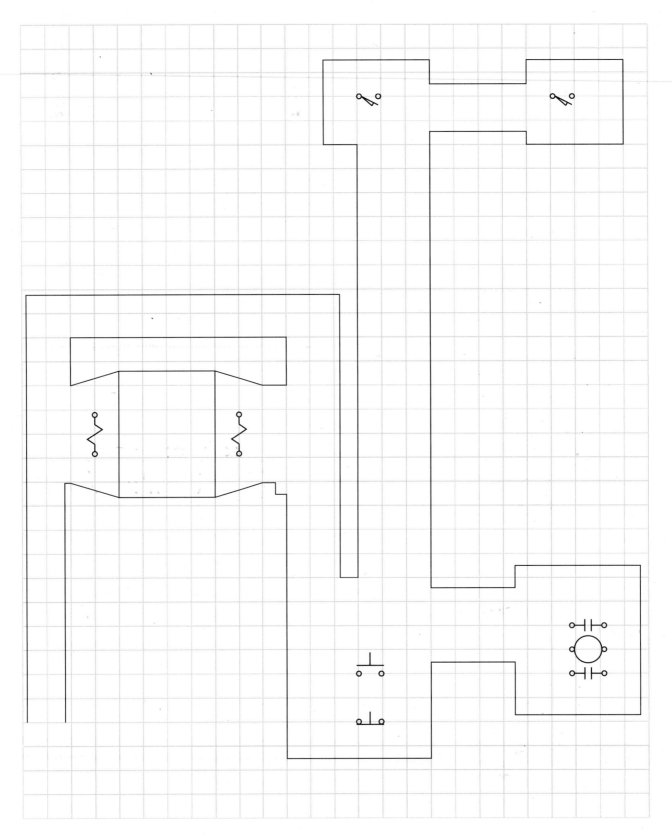

Time Delay and Logic

Name _____ Date _____

TECH-CHEK 7

Operational Diagrams

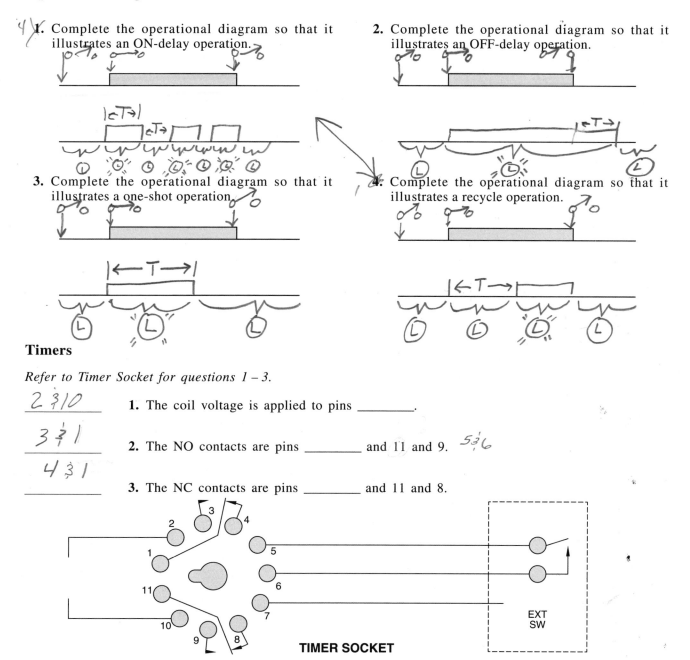

4 **1.** Complete the operational diagram so that it illustrates an ON-delay operation.

2. Complete the operational diagram so that it illustrates an OFF-delay operation.

3. Complete the operational diagram so that it illustrates a one-shot operation.

4. Complete the operational diagram so that it illustrates a recycle operation.

Timers

Refer to Timer Socket for questions 1 – 3.

2 3 10

3 2 1

4 3 1

1. The coil voltage is applied to pins _____.

2. The NO contacts are pins _____ and 11 and 9.   5 6

3. The NC contacts are pins _____ and 11 and 8.

EXT SW

TIMER SOCKET

Timing Codes

Add the timing code (X and O) for each load.

____X____   **1.** Load 1 timing code is __X__.

____O____   **2.** Load 2 timing code is __O__.

____O____   **3.** Load 3 timing code is __O__.

____X____   **4.** Load 4 timing code is __X__.

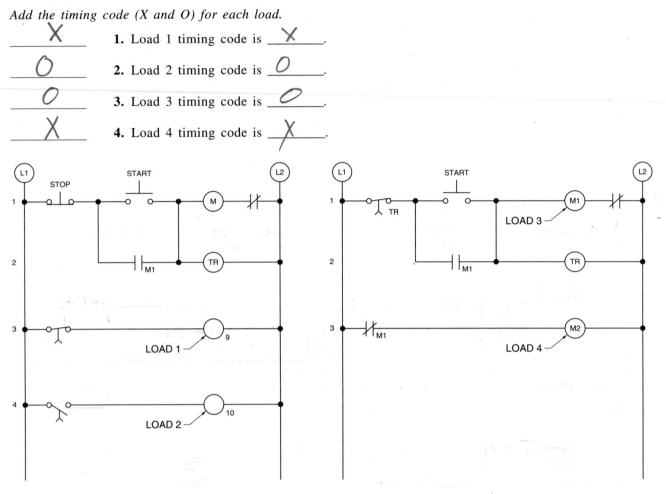

Timer Contact Symbols

1. Draw the symbol for an NO, timed-closed contact.

2. Draw the symbol for an NC, timed-open contact.

3. Draw the symbol for an NO, timed-open contact.

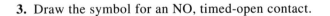

4. Draw the symbol for an NC, timed-closed contact.

Time Delay and Logic

Name _____ Date _____

WORKSHEET 7-1

Timer Settings

Use the timer information to set the DIP switches to the correct setting. Add line-reference, numerical cross-reference, wire-reference, and manufacturer's timer numbers to the line diagram.

1. The timer circuit provides for a two-hands, no tie-down operation. The operator has 2 sec to press each pushbutton. The circuit does not operate if any pushbutton is tied down or held in position.

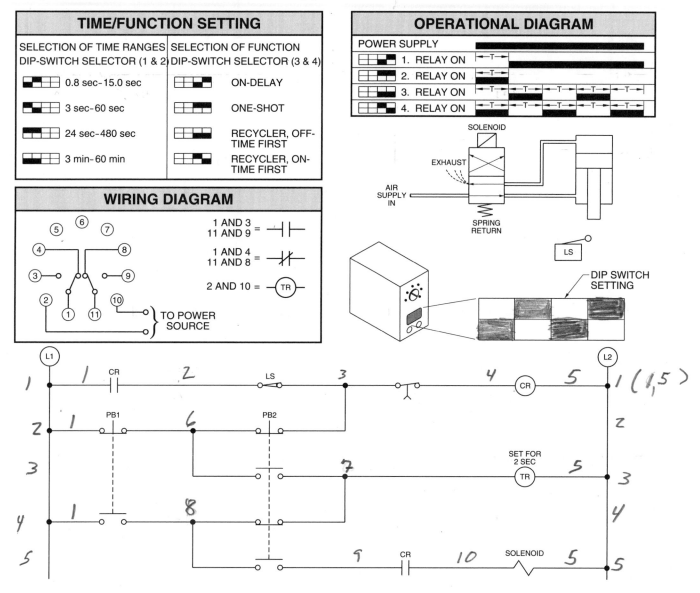

TIME/FUNCTION SETTING

SELECTION OF TIME RANGES DIP-SWITCH SELECTOR (1 & 2)	SELECTION OF FUNCTION DIP-SWITCH SELECTOR (3 & 4)
0.8 sec–15.0 sec	ON-DELAY
3 sec–60 sec	ONE-SHOT
24 sec–480 sec	RECYCLER, OFF-TIME FIRST
3 min–60 min	RECYCLER, ON-TIME FIRST

WIRING DIAGRAM

1 AND 3 / 11 AND 9 = —| |—
1 AND 4 / 11 AND 8 = —|/|—
2 AND 10 = —(TR)—

TO POWER SOURCE

OPERATIONAL DIAGRAM

POWER SUPPLY

	1. RELAY ON	
	2. RELAY ON	
	3. RELAY ON	
	4. RELAY ON	

SOLENOID

EXHAUST

AIR SUPPLY IN

SPRING RETURN

LS

DIP SWITCH SETTING

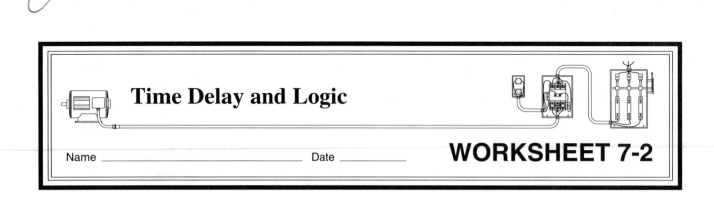

Time Delay and Logic

Name _____ Date _____

WORKSHEET 7-2

ON-Delay Motor Sequencing

Complete the wiring diagram for the line diagram. Do not make any wire splices or additional terminal connections on the wiring diagram. All connections must run from terminal screw to terminal screw.

1. The circuit is a start/stop pushbutton station with memory controlling two motors. A time-delay relay prevents both motors from starting simultaneously. An overload in Motor 1 shuts down the entire circuit. An overload in Motor 2 affects only Motor 2.

Time Delay and Logic

Name _____ Date _____

WORKSHEET 7-3

OFF-Delay Motor Control

Complete the wiring diagram for the line diagram. Do not make any wire splices or additional terminal connections on the wiring diagram. All connections must run from terminal screw to terminal screw.

1. In the circuit, Motor 2 starts and runs for a short time after Motor 1 has stopped. An overload in Motor 1 shuts down the entire circuit. An overload in Motor 2 affects only Motor 2.

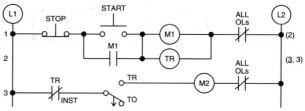

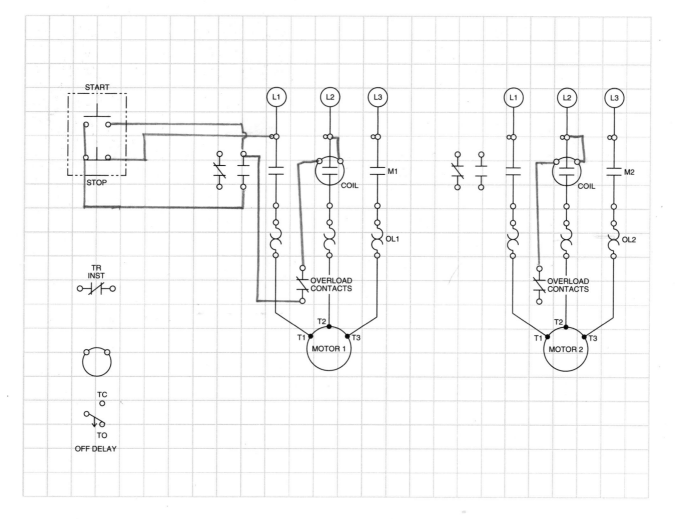

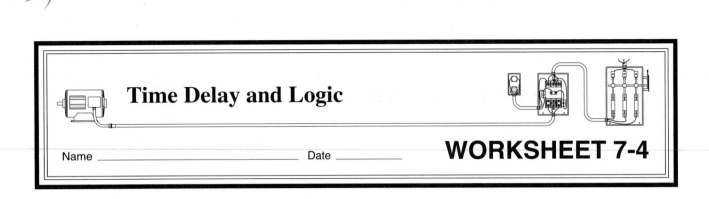

Time Delay and Logic

Name _____ Date _____

WORKSHEET 7-4

Timer Coding

Use Data Sheet D to complete the line diagram for the six circuits based on their established coding system for the load.

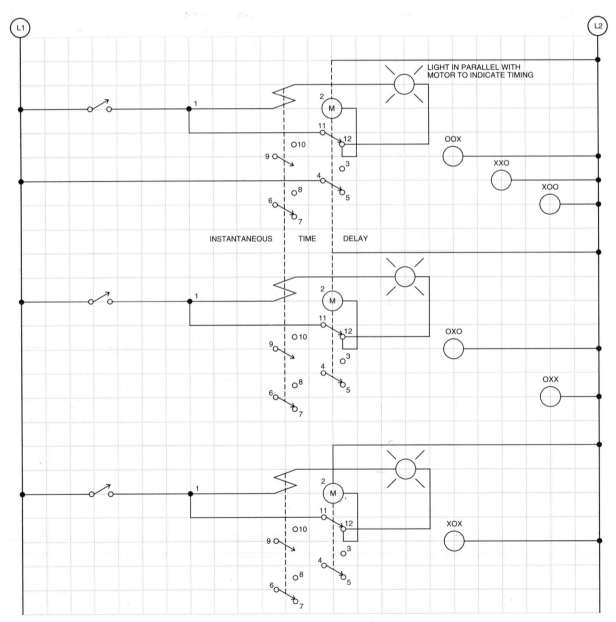

Time Delay and Logic

Name _____ Date _____

WORKSHEET 7-5

Timer Motor Sequencing

Use Data Sheet E to complete the timing circuit line diagram. Use standard lettering, numbering, and coding information.

1. A conveyor system is to be installed in which the first conveyor (controlled by M1) is turned ON by a standard start pushbutton with memory. After the first conveyor has run for 1 min, a second conveyor (controlled by M2) turns ON automatically. Both conveyors run until a standard stop pushbutton causes both to stop. Each conveyor motor should have independent overload protection.

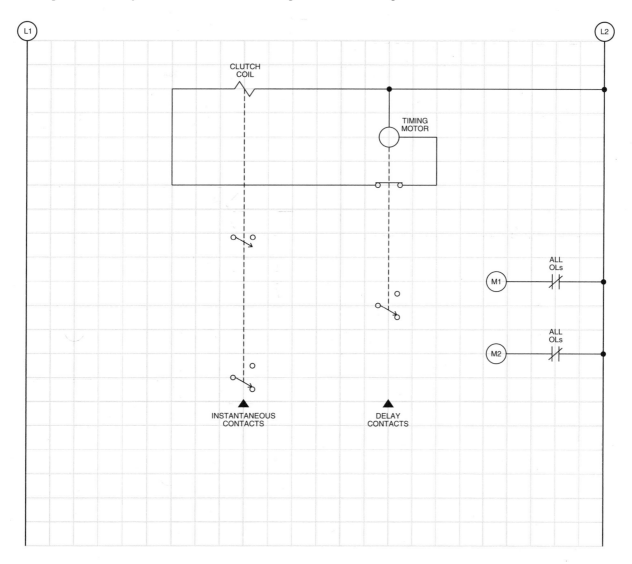

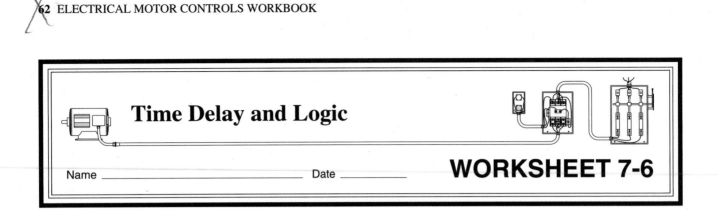

Time Delay and Logic

Name _____ Date _____

WORKSHEET 7-6

Timer Safety Circuit

Use Data Sheet E to complete the timing circuit line diagrams. Use standard lettering, numbering, and coding information.

1. A time control is to be installed to operate a sandblasting machine. A part is automatically sandblasted for 30 sec when the operator places the part in the machine, closes the door, and turns ON a toggle switch. The sandblaster is powered by a motor (controlled by M1). During sandblasting, a red pilot light illuminates, indicating danger. At all other times, a green pilot light illuminates, indicating it is safe to open the door. Overload protection should be provided for the motor.

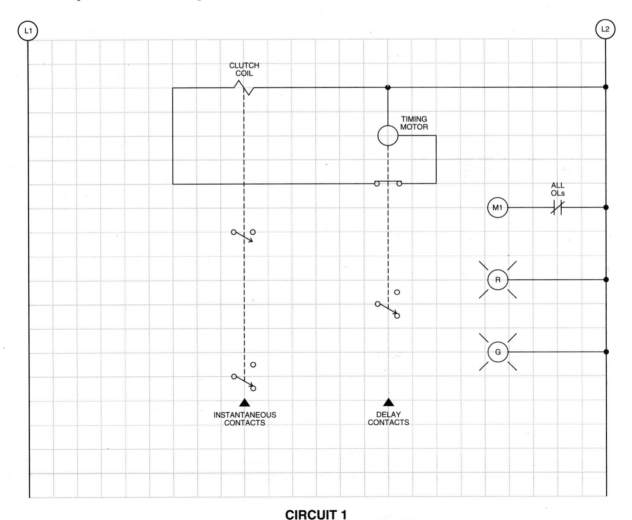

CIRCUIT 1

2. Redraw Circuit 1 so a limit switch may be built into the door of the machine to turn OFF the sandblasting operation any time the door is open.

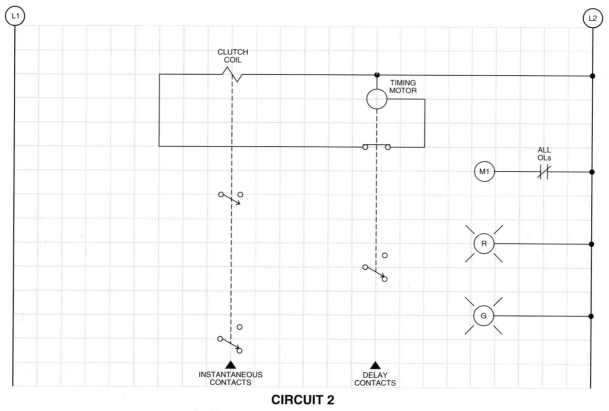

CIRCUIT 2

3. Redraw Circuit 2 so no sandblasting can take place unless the operator holds two pushbuttons down during the sandblasting operation.

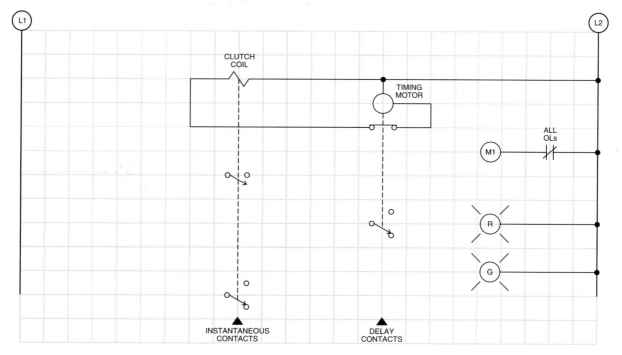

Time Delay and Logic

Name _____ Date _____

WORKSHEET 7-7

ON-Delay Timer Operational Diagrams

Complete the operational diagram for each load in the circuit.

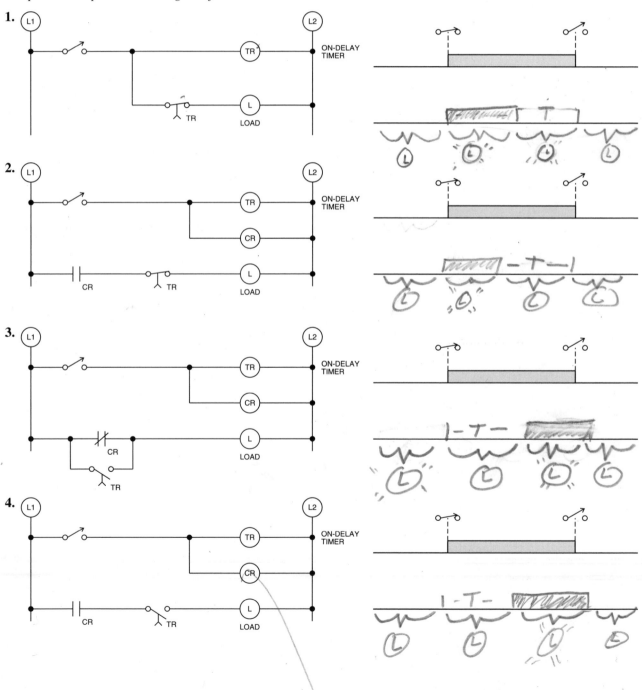

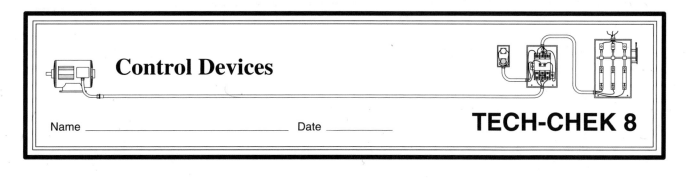

Control Devices

Name _____ Date _____

TECH-CHEK 8

Electrical Motor Controls

Manual. 1. A pushbutton is a(n) _____ control device.

auto 2. A pressure switch is a(n) _____ control device.

auto 3. A temperature switch is a(n) _____ control device.

auto 4. A flow switch is a(n) _____ control device.

auto 5. A liquid level switch is a(n) _____ control device.

auto 6. A limit switch is a(n) _____ control device.

B 7. In an electrical circuit, a limit switch is connected between _____.

    A. L2 and the load   C. L1 and L2
    B. L1 and the load   D. two loads

C 8. A _____ actuator is an actuator that has a long arm that may be cut to the required length.

    A. push-roller   C. wobble-stick
    B. standard   D. fork-lever

C 9. A _____ pressure sensing device is rated for 10,000 psi or more.

    A. bellows   C. piston
    B. diaphragm   D. spring

D 10. A _____ level switch is used to detect product dielectric variations.

    A. mechanical   C. conductive probe
    B. magnetic   D. capacitive

B 11. _____ loads are the least destructive loads to switch.

    A. Inductive   C. Capacitive
    B. Resistive   D. Motor

A 12. Most flow switches are designed to operate in the _____ position.

    A. horizontal   C. curved
    B. vertical   D. circular

B 13. A(n) _____ is used to control the yawing function of a windmill.

    A. anemometer   C. pressure switch
    B. wind vane   D. flow switch

A 14. A(n) _____ operator allows for easy emergency stops and operation with gloved hands.

    A. mushroom head   C. half-shrouded
    B. illuminated   D. flush

_C_   **15.** A _____ is a temperature-sensitive resistor that changes its electrical resistance with a change in temperature.

    A. bimetallic sensor   C. thermistor
    B. capillary tube   D. thermocouple

_C_   **16.** A(n) _____ is an output used to switch high-power DC loads.

    A. transistor   C. SCR
    B. triac   D. PTC

_A_   **17.** A(n) _____ is an output used to switch low-power DC loads.

    A. transistor   C. SCR
    B. triac   D. PTC

Closed   **18.** Pushbutton contacts are available as either normally open or normally _____.

Sump control   **19.** Charging and discharging tank applications are also known as pump and _____.

Smoke switch   **20.** A(n) _____ is a switch used in fire protection systems to detect burning material.

_NC_   **21.** The _____ contacts of a pressure switch are used when the pressure switch is used to maintain pressure in a tank.

Dead band   **22.** The _____ setting is the pressure range between the rising pressure and the falling pressure that is required to actuate the contacts.

Float   **23.** A(n) _____ control is used to maintain a small level differential in a tank.

Truth Table

Complete the truth table for the joystick circuit. Place an X in the appropriate box to indicate the closed contacts.

1.

POSITION	CONTACTS			
	A	B	C	D
RIGHT	X			
LEFT		X		
UP			X	
DOWN				X

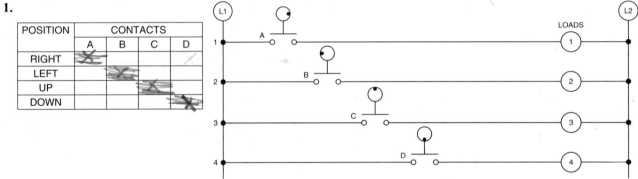

Limit Switch Connections

Determine whether the limit switches are correctly or incorrectly installed.

incorrectly   **1.** Limit Switch 1 is _____ installed.

correctly   **2.** Limit Switch 2 is _____ installed.

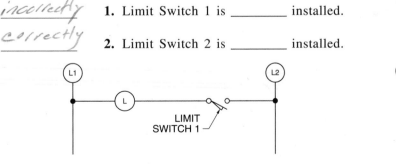
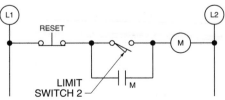

Control Devices

Name _____ Date _____

WORKSHEET 8-1

Truth Tables

Complete the truth tables for the circuits. Place an X in the appropriate box to indicate the closed contacts.

1.

POSITION	CONTACTS			
	A	B	C	D
AUTO	X	X	X	
HAND		X	X	X

2.

POSITION	CONTACTS			
	A	B	C	D
UP	X	X		
DOWN			X	
RIGHT				X

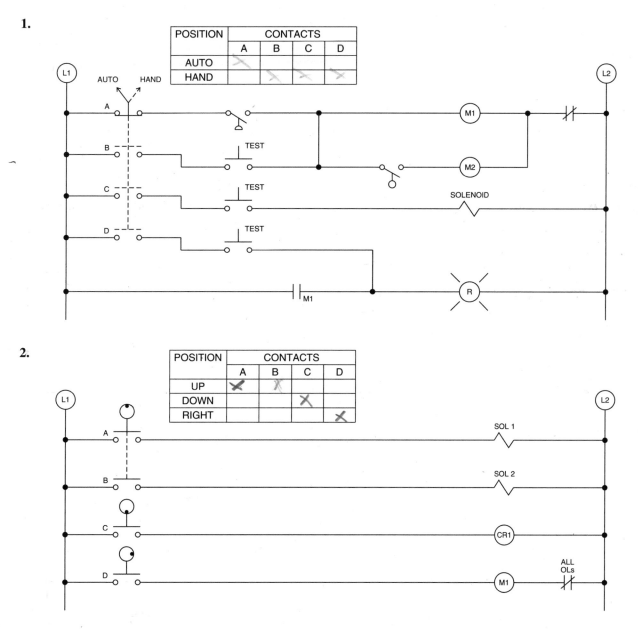

Control Devices

Name _____ Date _____

WORKSHEET 8-2

Heating Element Control

Complete the line and wiring diagrams according to the circuit information. Use standard lettering, numbering, and coding information.

1. Design a circuit in which a temperature switch is used to control a load even though the temperature switch contacts cannot directly handle the load. The heating contactor should energize at a low-temperature setting.

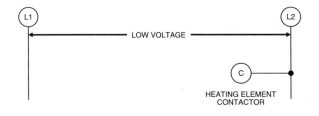

2. Complete the power circuit wiring diagram as a 3ϕ wye-connected set of heating elements.

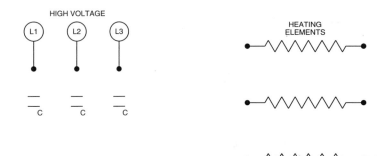

3. Complete the power circuit wiring diagram as a 3ϕ delta-connected set of heating elements.

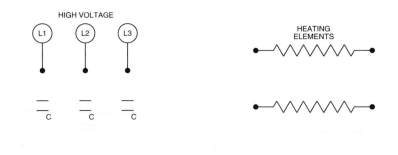

Control Devices

Name _____ Date _____

WORKSHEET 8-3

Vacuum Sensing

Complete the line diagrams according to the circuit information. Use standard lettering, numbering, and coding information.

1. Design a circuit with a vacuum switch that sounds a warning bell if a loss of vacuum occurs.

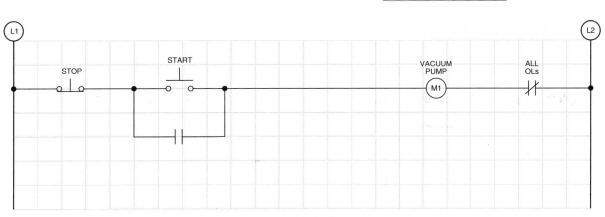

CIRCUIT 1

2. Redraw Circuit 1, adding a timer so the vacuum switch does not activate the bell unless the loss of vacuum occurs after the pump has run for 60 sec.

Control Devices

Name _____ Date _____

WORKSHEET 8-4

Pressure Control

Complete the line diagram based on the circuit information. Use standard lettering, numbering, and coding information.

1. Design a circuit with three pressure switches to maintain the proper amount of air pressure in an inflatable building. One pressure switch controls an air pump to keep the building inflated. A second switch detects overpressure that could rupture the building and warns of it by sounding a bell. The third switch detects underpressure that could cause the building to collapse and warns of it by sounding a horn.

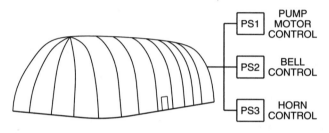

PS1 — PUMP MOTOR CONTROL

PS2 — BELL CONTROL

PS3 — HORN CONTROL

L1

L2

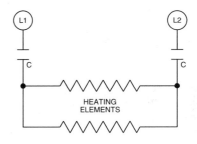

Control Devices

Name _____ Date _____

WORKSHEET 8-5

Dual-Temperature Control

Complete the line diagram based on the circuit information. Use standard lettering, numbering, and coding information.

1. Design a circuit with two separate temperature switches and a selector switch to provide two temperature controls. Heat is provided by heating elements activated through a magnetic contactor. The selector switch has three settings: high, low, and OFF. Temperature Switch 1 controls the high temperature and Temperature Switch 2 controls the low temperature. Complete the truth table for the selector switch to illustrate the circuit operation.

POSITION	CONTACTS	
	A	B
HIGH		
OFF		
LOW		

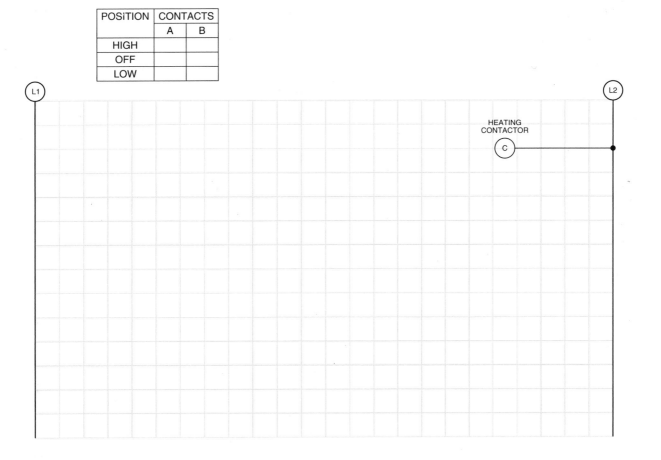

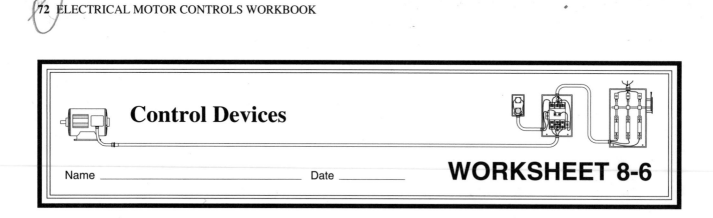

Fan/Heater Control

Complete the line diagram based on the circuit information. Use standard lettering, numbering, and coding information.

1. Design a circuit with a standard start/stop pushbutton station to control a fan motor and an electric heating contactor. Add a flow switch to ensure that the proper amount of air flow is present when the fan motor and heater are ON. A bell sounds if the flow is restricted when the fan is ON, but does not sound if the fan motor is OFF.

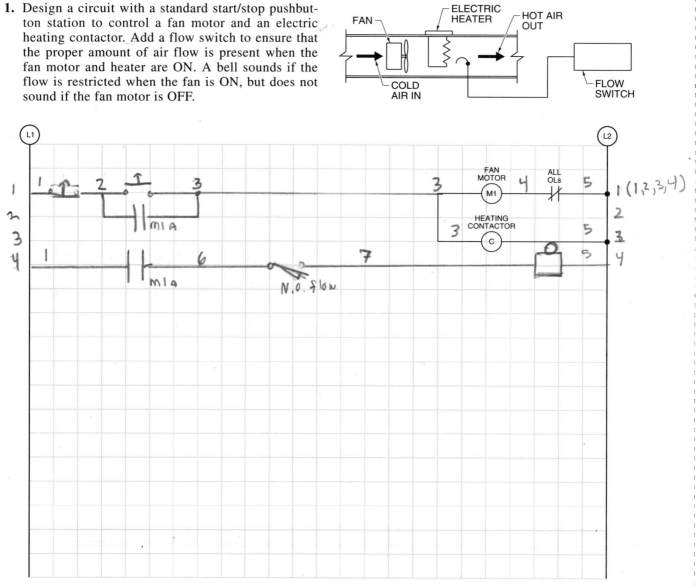

CIRCUIT 1

2. Redraw Circuit 1, adding a timer to keep the ventilation fan motor from operating for 30 sec after the heating element is turned ON. This allows a 30 sec warm-up period.

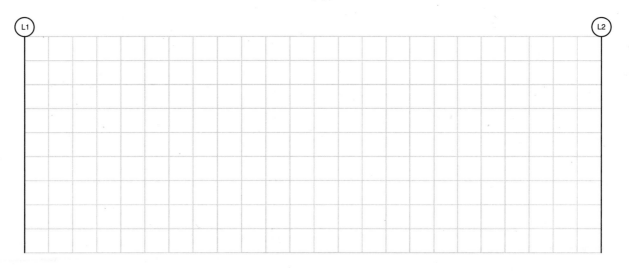

3. Redraw Circuit 1, changing the control from a pushbutton control circuit to an automatic temperature control circuit. Include a 2-position selector switch that can be placed in an automatic position or an OFF position. In the automatic position, the temperature switch maintains the correct temperature. In the OFF position, the heating element and fan motor cannot operate.

Control Devices

Name _____ Date _____

WORKSHEET 8-7

Level Control Relay

Use Data Sheet F to complete the line diagram. Use standard lettering, numbering, and coding information.

1. Design a circuit where the relay is connected to control a pump motor. The relay is connected to monitor two levels of fluid. The pump motor turns ON when the maximum level setting is reached, and stays ON until the minimum level setting is reached. Indicate only those connections necessary to form the control circuit.

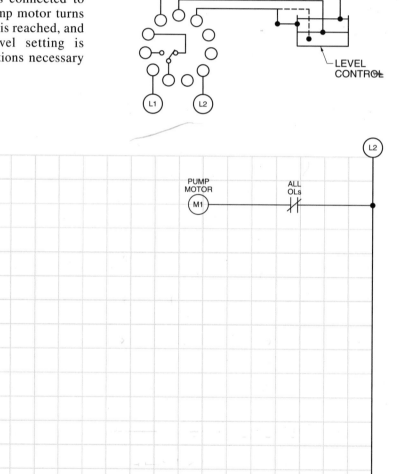

LEVEL CONTROL

L1

L2

PUMP MOTOR

M1

ALL OLs

Reversing Motor Circuits

Name _____ Date _____

TECH-CHEK 9

Electrical Motor Controls

L3

1. The industry standard is to interchange power leads L1 and _____ to reverse the direction of rotation of a 3φ motor.

Centrifugal Switch

2. A(n) _____ is a device used to disconnect the starting winding of a 1φ motor.

Starting

3. The industry standard is to interchange the leads of the _____ winding to reverse the direction of rotation of a 1φ motor.

Running

4. The _____ winding of a 1φ motor normally has the lowest resistance.

Capacitor

5. A(n) _____ is added to a 1φ motor to develop more starting torque.

Current

6. A motor is connected for a high voltage so the _____ is reduced.

armature

7. The industrial standard for reversing a DC series motor is to reverse the current through the _____.

10

8. When troubleshooting a reversing circuit, the voltage must be within _____% of the control circuit's rating.

Jog

9. A(n) _____ circuit allows the operator to start the motor for a short time without memory.

Overloads

10. A drum switch is not considered a motor starter because the switch does not contain _____.

Push button
P/L starter

11. The three types of interlocking used in Circuit 1 are mechanical, auxiliary contact, and _____.

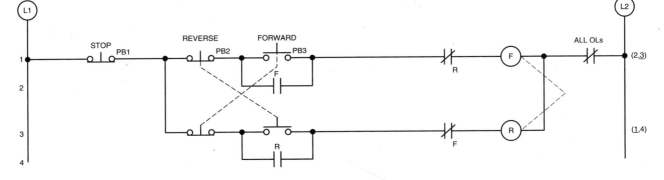

CIRCUIT 1

Motor Leads

1. Mark each motor lead as T1, T2, etc.

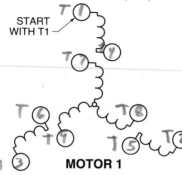

START WITH T1

MOTOR 1

2. Mark each motor lead as T1, T2, etc.

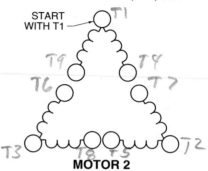

START WITH T1

MOTOR 2

3. Mark each motor lead as A1, A2, S1, S2, etc.

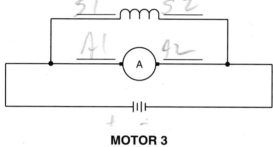

MOTOR 3

4. Mark each motor lead as A1, A2, S1, S2, etc.

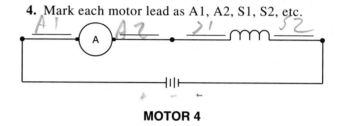

MOTOR 4

5. Mark each motor lead as A1, A2, S1, S2, etc.

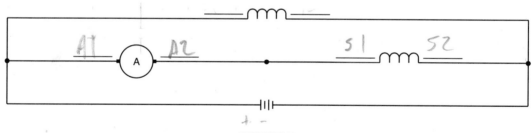

MOTOR 5

Motor Identification

Wye   **1.** Motor 1 is a(n) _____ motor.

Delta   **2.** Motor 2 is a(n) _____ motor.

DC shunt   **3.** Motor 3 is a(n) _____ motor.

DC series   **4.** Motor 4 is a(n) _____ motor.

DC compound   **5.** Motor 5 is a(n) _____ motor.

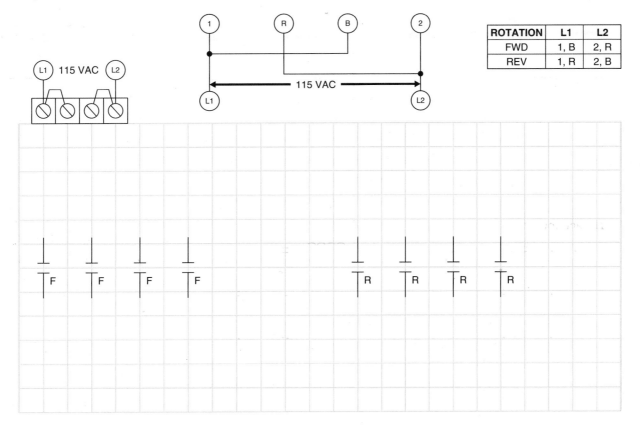

Reversing Motor Circuits

Name _____   Date _____

WORKSHEET 9-1

Reversing 1φ Motors

Use Data Sheet G to draw the wiring diagram. Note: *Use only the number of contacts required.*

1. Operate the motor in forward and reverse at 115 VAC.

ROTATION	L1	L2
FWD	1, B	2, R
REV	1, R	2, B

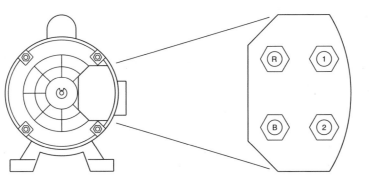

Reversing Motor Circuits

Name _____ Date _____

WORKSHEET 9-2

Reversing 3φ Motors

Use Data Sheet G to draw the wiring diagram. Note: Interchange any two power leads to reverse the rotation of the motor. The accepted standard is to interchange leads T1 and T3. Use only the number of contacts required.

1. Operate the motor in forward and reverse at 230 VAC.

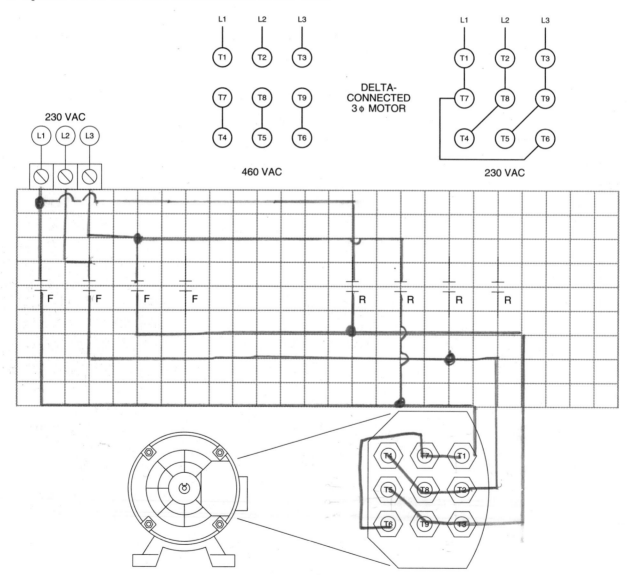

Reversing Motor Circuits

Name _____ Date _____

WORKSHEET 9-3

Reversing Dual-Voltage, 1φ Motors at Low Voltage

Use Data Sheet G to draw the wiring diagram. Note: *Interchange the red and black leads to reverse the motor rotation. Use only the number of contacts required.*

1. Operate the motor in forward and reverse at 120 VAC.

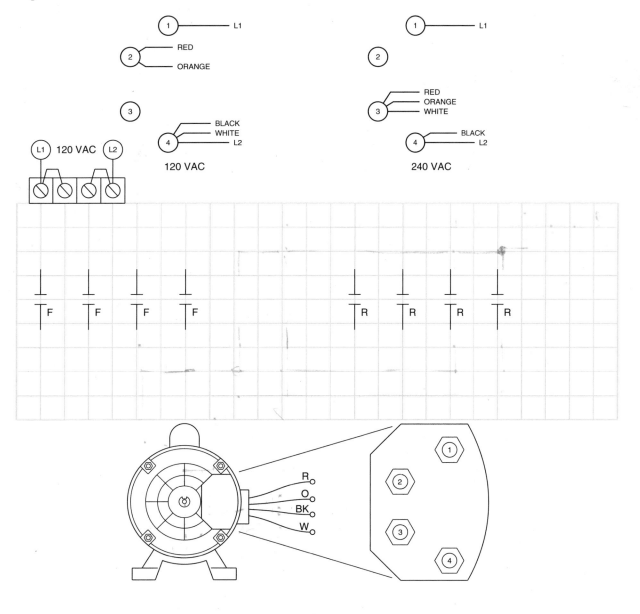

Reversing Motor Circuits

Name _____ Date _____

Reversing Dual-Voltage, 1ϕ Motors at High Voltage

Use Data Sheet G to draw the wiring diagram. Note: Interchange the red and black leads to reverse the motor rotation. Use only the number of contacts required.

1. Operate the motor in forward and reverse at 230 VAC.

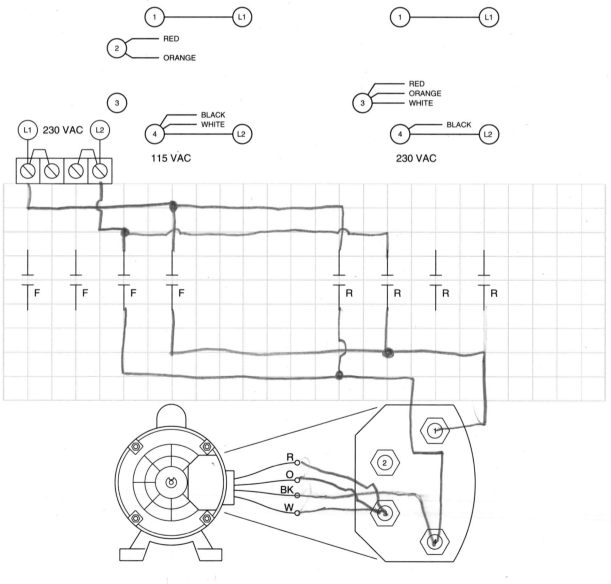

Reversing Motor Circuits

Name _____ Date _____

WORKSHEET 9-5

Reversing 1φ Motors – All Wires Not Used

Use Data Sheet G to draw the wiring diagram. Note: Interchange the red and black leads to reverse the motor rotation. Use only the number of contacts required.

1. Operate the motor in forward and reverse at 230 VAC.

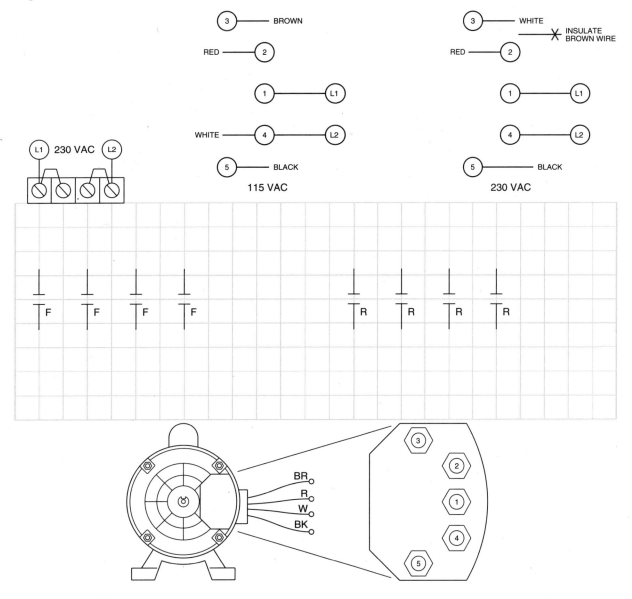

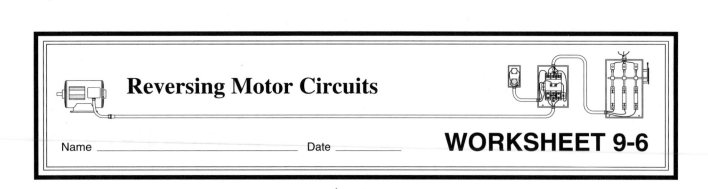

Reversing Motor Circuits

Name _____ Date _____

WORKSHEET 9-6

Interchanging Motor Leads

Use Data Sheet G to draw the wiring diagram. Note: Interchange the red and black leads to reverse the motor rotation. Use only the number of contacts required.

1. Operate the motor in forward and reverse at 115 VAC.

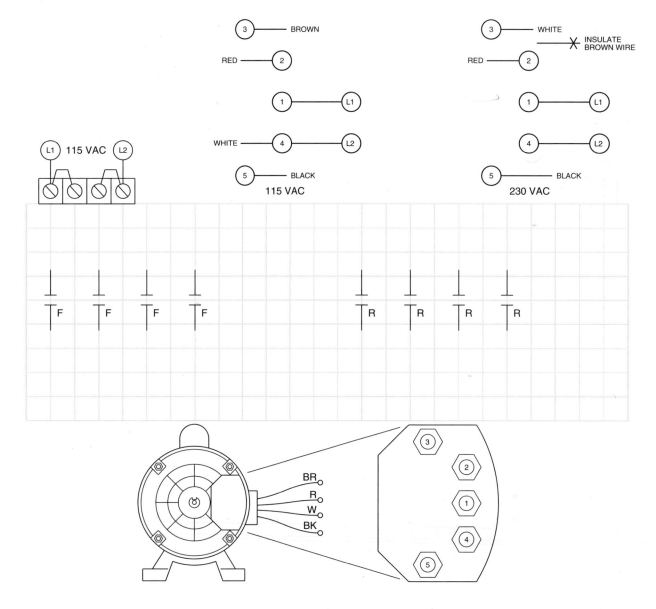

Reversing Motor Circuits

Name _____ Date _____

WORKSHEET 9-7

Reversing 2-Speed, 1φ Motors at Low Speed

Use Data Sheet G to draw the wiring diagram. Note: Interchange the red and black leads to reverse the motor rotation. Use only the number of contacts required.

1. Operate the motor in forward and reverse at low speed.

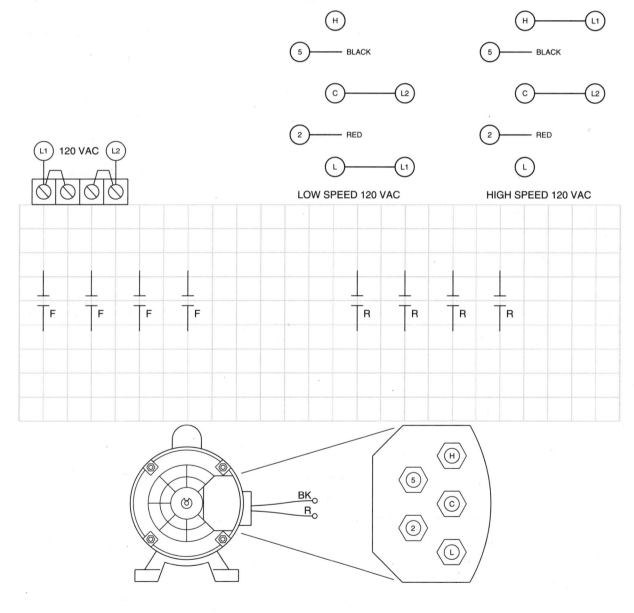

LOW SPEED 120 VAC   HIGH SPEED 120 VAC

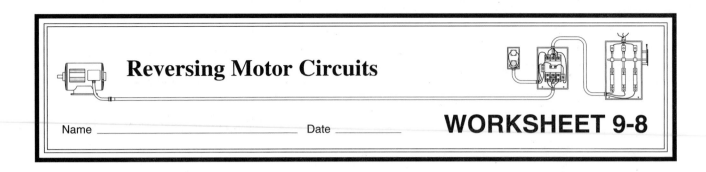

Reversing Motor Circuits

Name _____ Date _____

WORKSHEET 9-8

Reversing 2-Speed, 1φ Motors at High Speed

Use Data Sheet G to draw the wiring diagram. Note: *Interchange the red and black leads to reverse the motor rotation. Use only the number of contacts required.*

1. Operate the motor in forward and reverse at high speed.

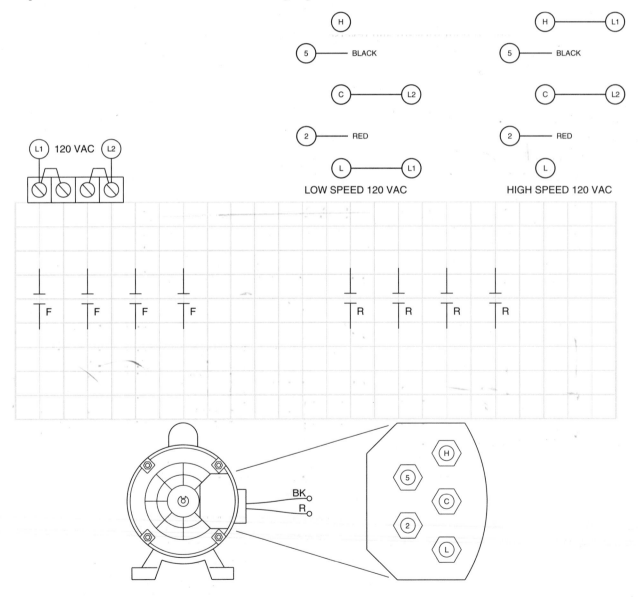

Reversing Motor Circuits

WORKSHEET 9-9

Name _____ Date _____

Limit Switch Motor Stopping

Complete the wiring diagram based on the line diagram. Do not make any wire splices or additional terminal connections on the wiring diagram. All connections must run from terminal screw to terminal screw.

1. The line diagram is of a standard forward/reverse/stop pushbutton station for forwarding and reversing a motor. Included in the circuit are mechanical and auxiliary contact interlocking. Also included are a forward limit switch to stop the motor in forward and a reverse limit switch to stop the motor in reverse. Overload protection is common to both forward and reverse directions.

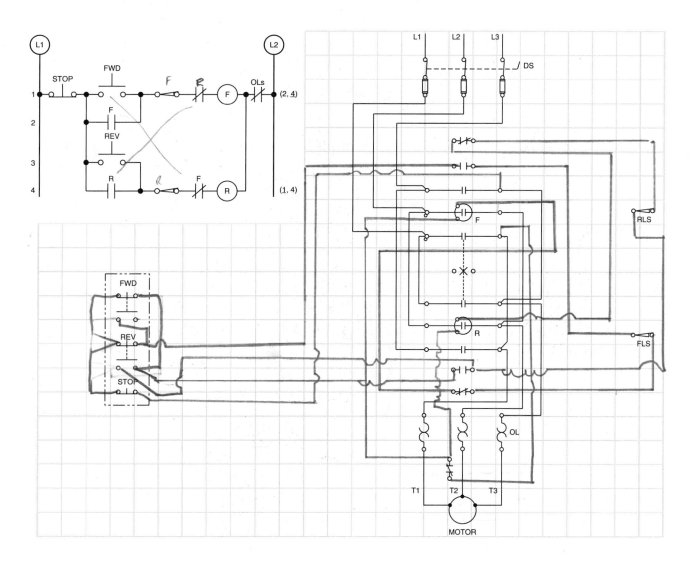

Reversing Motor Circuits

Name _____ Date _____

WORKSHEET 9-10

Forward/Reverse Circuit with Indicator Lights

Complete the wiring diagram based on the line diagram. Do not make any wire splices or additional terminal connections on the wiring diagram. All connections must run from terminal screw to terminal screw.

1. The line diagram is of a standard forward/reverse/stop pushbutton station with indicating lights to show the direction of motor rotation. The indicating lights are to be mounted within the pushbutton enclosure. Overload protection is common to both forward and reverse directions.

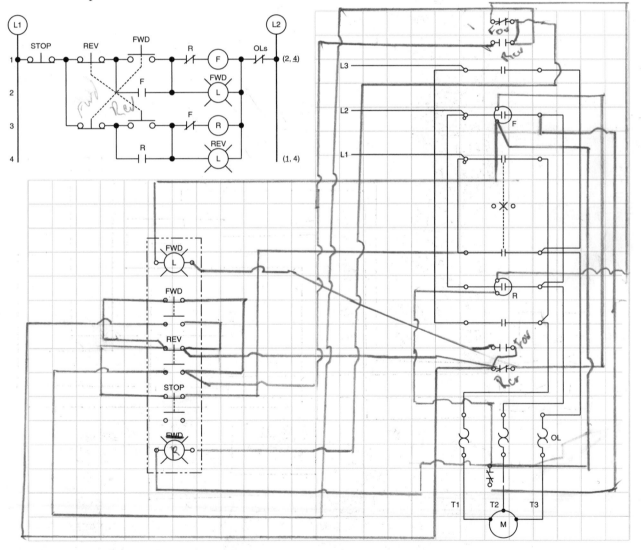

Reversing Motor Circuits

Name _____ Date _____ **WORKSHEET 9-11**

Selector Switch Motor Control

Complete the wiring diagram based on the line diagram. Do not make any wire splices or additional terminal connections on the wiring diagram. All connections must run from terminal screw to terminal screw.

1. The line diagram is of a standard start/stop pushbutton station with a selector switch to control direction of motor travel. A visual indication of the direction of motor rotation is provided on the selector switch in case the motor and drive unit cannot be seen from the control station. Overload protection is common to forward and reverse directions.

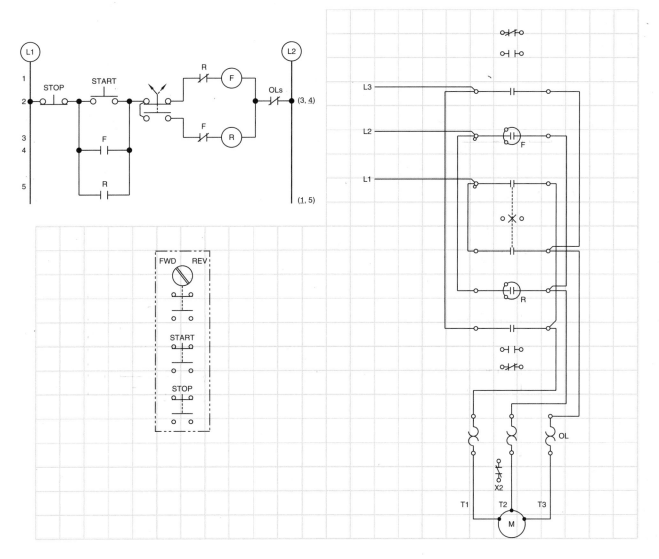

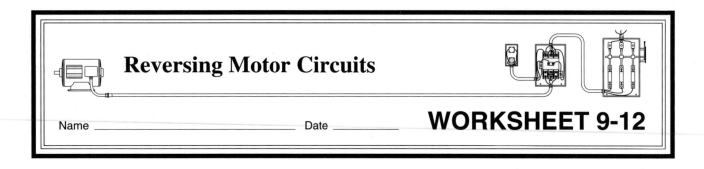

Reversing Motor Circuits

Name _____ Date _____

WORKSHEET 9-12

Selector Switch Jog/Run Control

Complete the wiring diagram based on the line diagram. Do not make any wire splices or additional terminal connections on the wiring diagram. All connections must run from terminal screw to terminal screw.

1. The line diagram is of a standard forward/reverse/stop pushbutton station with a selector switch to provide for jogging or running. The forward and reverse pushbuttons energize the motor only as long as they are pressed when the selector switch is in the jog position. The forward and reverse pushbuttons operate as a standard pushbutton station with memory when the selector switch is in the run position. Overload protection is common to both forward and reverse directions.

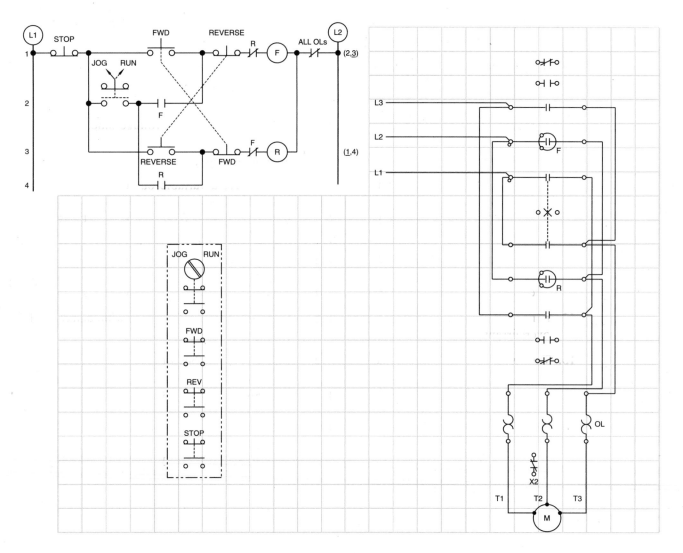

Power Distribution Systems

Name _____ Date _____

TECH-CHEK 10

Electrical Motor Controls

_____ **1.** A(n) _____ is the major source of electrical power.

_____ **2.** A(n) _____ connection is made when the three ends of separate phases are connected together, creating a common wire.

_____ **3.** A(n) _____ connection is made when the end of one separate phase is connected to the beginning of the next phase, etc.

_____ **4.** In a transformer, the _____ winding is the coil that draws power from the source.

_____ **5.** In a transformer, the _____ winding is the coil that delivers the energy at a transformed or changed voltage to a load.

_____ **6.** Step-up or step-down, when used with transformers, always refers to the _____.

_____ **7.** In a transformer, a marking of X1 or X2 indicates the _____ voltage side.
  A. high
  B. low
  C. primary
  D. neither A, B, nor C

_____ **8.** A transformer is provided with _____ to provide for a variable output.

_____ **9.** As far as the power company is concerned, the _____ is the last point on the power distribution system.

_____ **10.** Switchboards are designed for use as distribution, _____, or a combination of both.

_____ **11.** The three types of panelboards are power, distribution, and _____ panelboards.

_____ **12.** A(n) _____ is a reusable OCPD that opens a circuit automatically at a predetermined overcurrent.

_____ **13.** The two basic types of bus ducts are feeder and _____ bus ducts.

_____ **14.** _____ is the process of delivering electrical power where it is needed.

_____ **15.** A(n) _____ is a circuit that converts AC to DC.

_____ **16.** _____ are used on an alternator so that the external load may be easily attached to the rotor.

_____ **17.** A(n) _____-connected system is regarded as the safest of all distribution systems.

_____ **18.** The capacity of transformers is rated in _____.

_____ **19.** The three main sections of a substation are the primary switchgear, transformer, and _____ sections.

_____ **20.** Switchboards are rated as to the maximum voltage and _____ they can handle.

_____ **21.** A(n) _____ circuit is the portion of a distribution system between the final overcurrent protection device and the outlet or load connected to it.

_____ **22.** A(n) _____ electrode is a conductor embedded in the earth to provide a good ground.

_____ **23.** _____ perform the same function as fuses and are tested the same way.

_____ **24.** A(n) _____ is an electrical interface designed to change AC from one voltage level to another.

_____ **25.** The function of a substation includes _____.

    A. receiving voltage and increasing   C. providing a place to adjust and
       it to an appropriate level   regulate outgoing voltage
    B. providing a safe point for dis-   D. A, B, and C
       connecting power

_____ **26.** A(n) _____ is the piece of equipment in which a large block of electric power is delivered from a substation and broken down into smaller blocks for distribution throughout a building.

_____ **27.** A(n) _____ is a wall-mounted distribution cabinet containing a group of overcurrent and short-circuit protection devices for lighting, appliance, or power distribution branch circuits.

_____ **28.** A(n) _____ is a metal-enclosed distribution system of bus bars available in pre-fabricated sections.

_____ **29.** A(n) _____ receives incoming power and delivers it to the control circuit and motor loads.

_____ **30.** A(n) _____ is a device which protects transformers and other electrical equipment from voltage surges caused by lightning.

Transformers

_____ **1.** The primary side voltage is _____ V.

_____ **2.** The primary side current is _____ A.

_____ **3.** The secondary side current is _____ A.

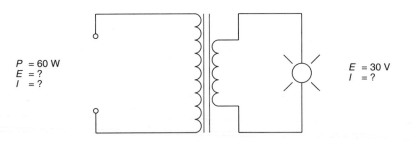

$P = 60$ W
$E = ?$
$I = ?$

$E = 30$ V
$I = ?$

TRANSFORMER RATIO = 4 TO 1

4 - Ind Elec II

Power Distribution Systems

Name _____ Date _____ **WORKSHEET 10-1**

Delta-to-Delta Connections

Complete the transformer wiring diagram in a delta-to-delta transformer bank connection.

1. Connect the primary transformer lines to the distribution system to form a delta-connected primary. Connect the secondary transformer lines to the distribution system to provide 3ϕ, high-voltage 1ϕ, and low-voltage 1ϕ power. Connect each load to the correct power supply.

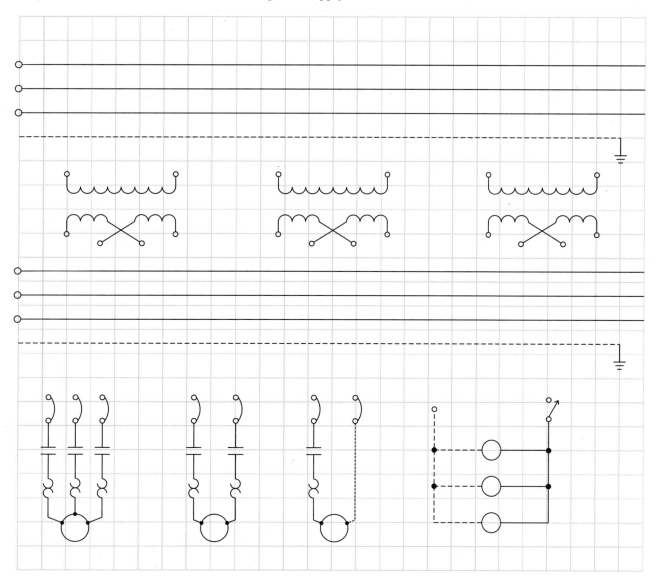

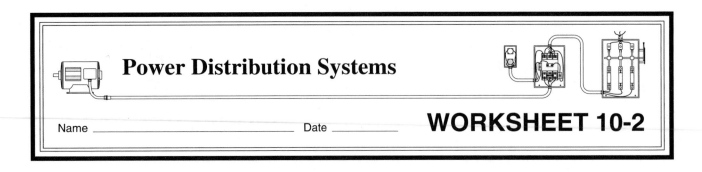

Power Distribution Systems

Name _____ Date _____

WORKSHEET 10-2

Wye-to-Wye Connections

Complete the transformer wiring diagram in a wye-to-wye transformer bank connection.

1. Connect the primary transformer lines to the distribution system to form a wye-connected primary. Connect the secondary transformer lines to the distribution system to form a wye-connected secondary that provides 3φ, high-voltage 1φ, and low-voltage 1φ power. Connect each load to the correct power supply.

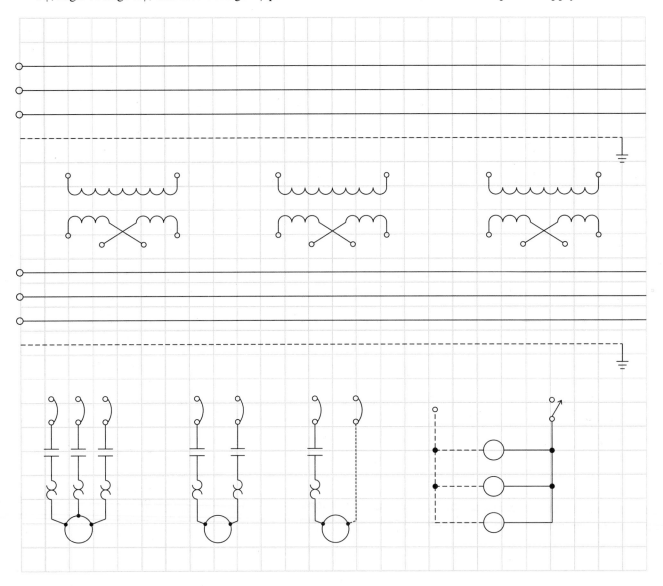

Power Distribution Systems

Name _____ Date _____ **WORKSHEET 10-3**

Delta-to-Wye Connections

Complete the transformer wiring diagram in a delta-to-wye transformer bank connection.

1. Connect the primary transformer lines to the distribution system to form a delta-connected primary. Connect the secondary transformer lines to the distribution system to form a wye-connected secondary that provides 3ϕ, high-voltage 1ϕ, and low-voltage 1ϕ power. Connect each load to the correct power supply.

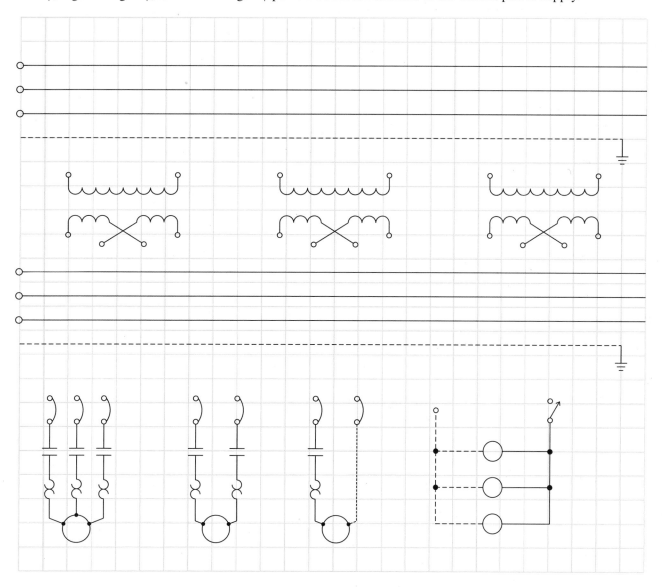

Power Distribution Systems

Name _____ Date _____ **WORKSHEET 10-4**

Wye-to-Delta Connections

Complete the transformer wiring diagram in a wye-to-delta transformer bank connection.

1. Connect the primary transformer lines to the distribution system to form a delta-connected secondary that provides 3φ, high-voltage 1φ, and low-voltage 1φ power. Connect each load to the correct power supply.

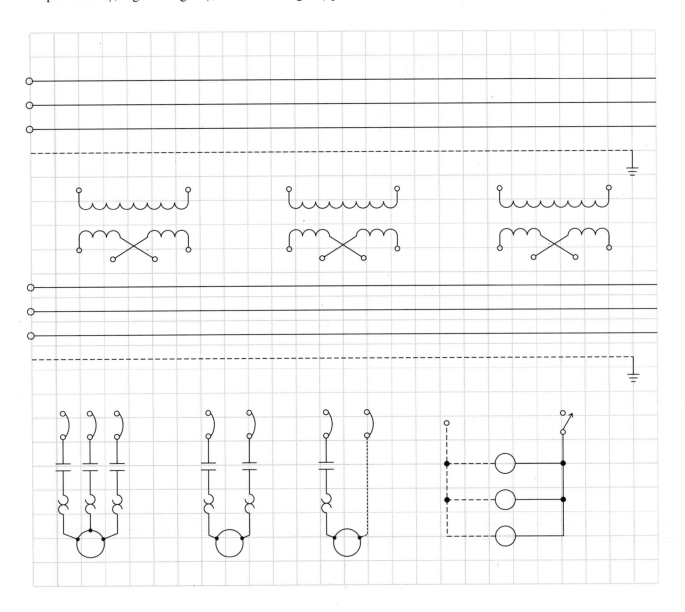

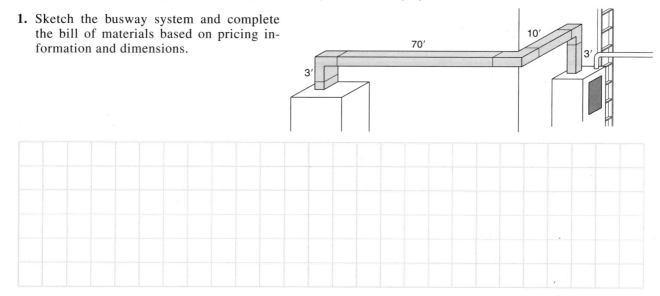

Power Distribution Systems

Name _____ Date _____

WORKSHEET 10-5

Busway System Design

Use Data Sheet H to complete the bill of materials for the busway system.

1. Sketch the busway system and complete the bill of materials based on pricing information and dimensions.

PRICING INFORMATION (LIST)		
Feeder busway duct	$40.00	per ft
Plug-in busway duct	45.00	per ft
Elbows	170.00	ea
Crosses	225.00	ea
50 A breakers	175.00	ea
100 A breakers	225.00	ea

BILL OF MATERIALS

Totals

_____ ft of feeder duct @ $_____ per ft $_____

_____ ft of plug-in duct @ $_____ per ft   _____

_____ elbows @ $_____ ea   _____

_____ tees @ $_____ ea   _____

_____ crosses @ $_____ ea   _____

_____ 50 A breakers @ $_____ ea   _____

_____ 100 A breakers @ $_____ ea   _____

SUBTOTAL _____ (LIST)

– 20% DISCOUNT _____

TOTAL _____

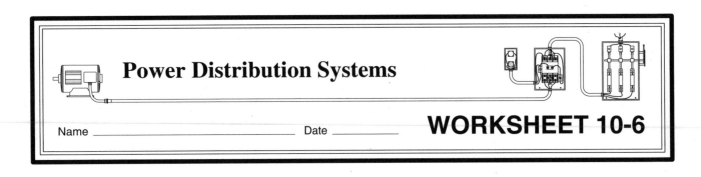

Power Distribution Systems

Name _____ Date _____

WORKSHEET 10-6

Busway System Calculation

Complete the bill of materials for the busway system.

1. Use the pricing information and dimensions to complete the bill of materials.

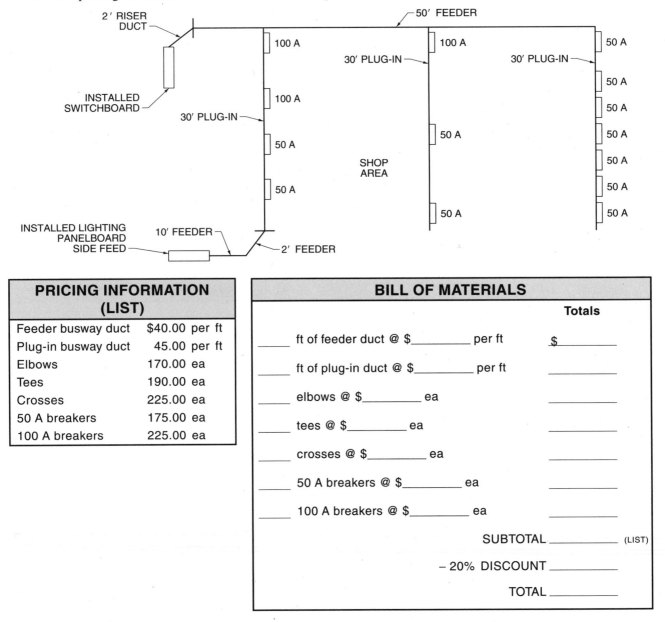

PRICING INFORMATION (LIST)	
Feeder busway duct	$40.00 per ft
Plug-in busway duct	45.00 per ft
Elbows	170.00 ea
Tees	190.00 ea
Crosses	225.00 ea
50 A breakers	175.00 ea
100 A breakers	225.00 ea

BILL OF MATERIALS

Totals

_____ ft of feeder duct @ $_____ per ft $_____

_____ ft of plug-in duct @ $_____ per ft   _____

_____ elbows @ $_____ ea   _____

_____ tees @ $_____ ea   _____

_____ crosses @ $_____ ea   _____

_____ 50 A breakers @ $_____ ea   _____

_____ 100 A breakers @ $_____ ea   _____

SUBTOTAL _____ (LIST)

– 20% DISCOUNT _____

TOTAL _____

Power Distribution Systems

Name _____ Date _____

WORKSHEET 10-7

Wiring Devices

Complete the current, voltage, and phase rating for each receptacle configuration.

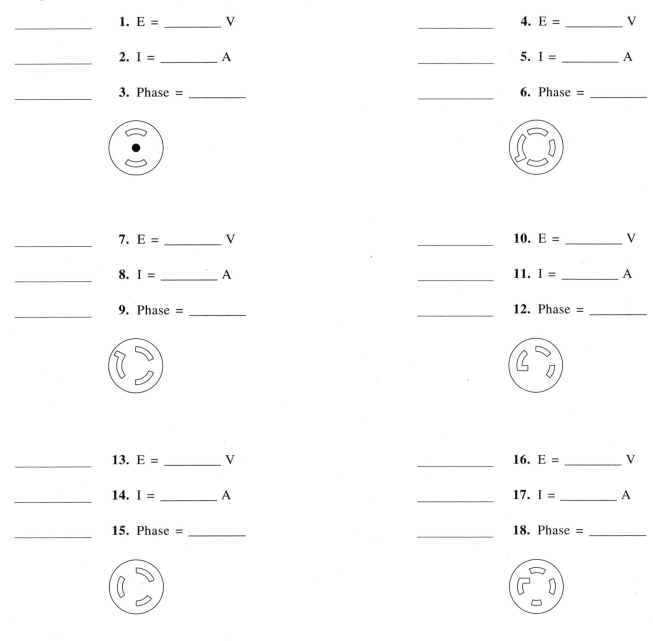

_____ 1. E = _____ V

_____ 2. I = _____ A

_____ 3. Phase = _____

_____ 4. E = _____ V

_____ 5. I = _____ A

_____ 6. Phase = _____

_____ 7. E = _____ V

_____ 8. I = _____ A

_____ 9. Phase = _____

_____ 10. E = _____ V

_____ 11. I = _____ A

_____ 12. Phase = _____

_____ 13. E = _____ V

_____ 14. I = _____ A

_____ 15. Phase = _____

_____ 16. E = _____ V

_____ 17. I = _____ A

_____ 18. Phase = _____

Power Distribution Systems

Name _____ Date _____

WORKSHEET 10-8

Start/Stop Circuit Control

Complete the wiring diagram according to the line diagram. Do not make any wire splices or additional terminal connections on the wiring diagram. All connections must run from terminal screw to terminal screw.

1. Wire the blank motor control panel as a standard start/stop pushbutton control with memory and an indicating light. Connect the wiring of the motor control panel to the terminal blocks. Do not connect the pushbutton or indicator light directly to the terminal blocks. Use blank spaces adjoining the pushbutton and indicator light to indicate by number which terminals would be connected to these points. For example, the stop pushbutton is marked with wire numbers 1 and 2.

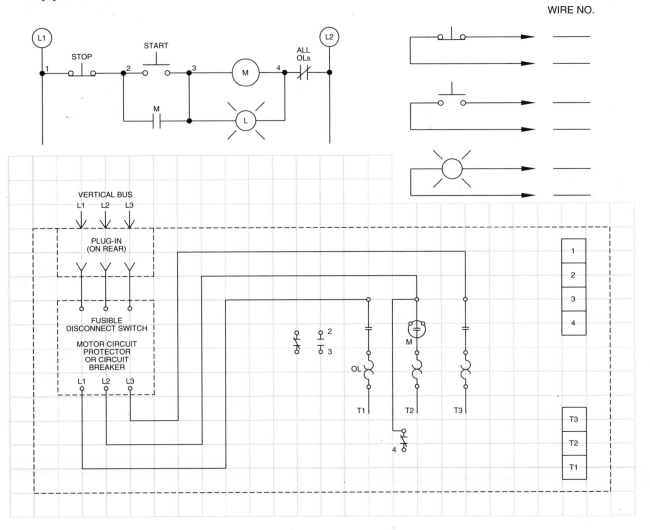

Power Distribution Systems

Name _____ Date _____

Jogging Circuit Control

Complete the wiring diagram according to the line diagram. Do not make any wire splices or additional terminal connections on the wiring diagram. All connections must run from terminal screw to terminal screw.

1. Wire the blank motor control panel as a jogging circuit with a control relay. Connect the wiring of the motor control panel to the terminal blocks provided. Do not connect control devices directly to the terminal blocks. Use blank spaces adjoining the control devices to indicate by number which terminals would be connected to these points.

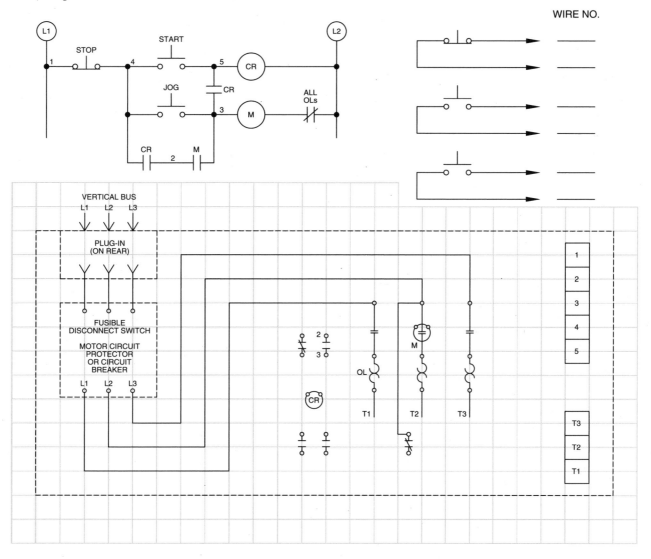

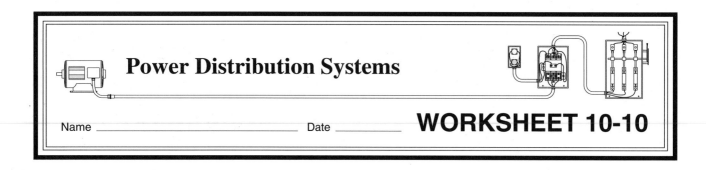

Power Distribution Systems

Name _____ Date _____

WORKSHEET 10-10

Surge and Backspin Protection

Complete the wiring diagram according to the line diagram. Do not make any wire splices or additional terminal connections on the wiring diagram. All connections must run from terminal screw to terminal screw.

1. Wire the blank motor control panel as a circuit that provides surge and backspin protection by means of a time-delay relay. Do not connect control devices directly to the terminal blocks. Use blank spaces adjoining control devices to indicate by number which terminals would be connected to these points.

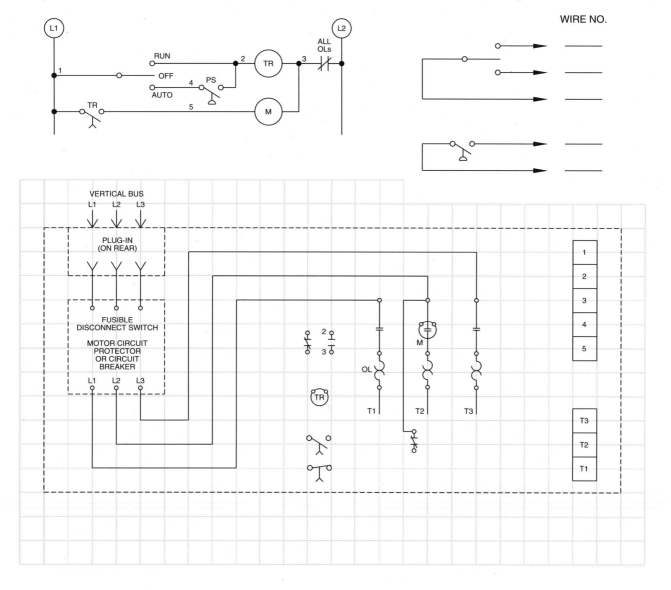

Solid-State Electronic Control Devices

Name _____ Date _____

TECH-CHEK 11

Electrical Motor Controls

_____ **1.** _____ on a PC board are small, round conductors to which component leads are soldered.

_____ **2.** _____ are used to interconnect two or more pads.

_____ **3.** A(n) _____ is a PC board with multiple terminations on one end.

_____ **4.** The _____ is the central core of an atom.

_____ **5.** _____ is the process by which the crystal structure of an atom is altered.

_____ **6.** The _____ is the junction of P-type and N-type materials in a diode.

_____ **7.** A(n) _____ is a thermally-sensitive resistor.

_____ **8.** A(n) _____ converts light energy into electrical energy.

_____ **9.** A(n) _____ is a transducer that changes resistance with a corresponding change in pressure.

_____ **10.** _____ is the ratio of the amplitude of an output signal to the amplitude of the input signal.

_____ **11.** _____ is the maximum reverse bias voltage of a diode.

    A. Forward current   C. Peak inverse voltage
    B. Reverse current   D. Depletion region

_____ **12.** A(n) _____ is used to provide conduction from several sources on a PC board.

    A. edge card   C. trace
    B. bus   D. pad

_____ **13.** _____ are free electrons in any conductor.

    A. Neutrons   C. Foils
    B. Carriers   D. Diodes

_____ **14.** _____ current is the current value that keeps an SCR ON as long as the current stays above that value.

    A. Avalanche   C. Breakover
    B. Blocking   D. Holding

_____ **15.** A(n) _____ provides a complete circuit function in one semiconductor package.

    A. triac   C. integrated circuit
    B. diac   D. breakover diac

_____ 16. _____ electrons are the electrons in the outermost shell of an atom.

_____ 17. _____ are the missing electrons in the crystal structure.

_____ 18. _____ devices are devices in which electrical conductivity is between that of a conductor and that of an insulator.

_____ 19. _____ are electronic components that allow current to pass through them in only one direction.

_____ 20. A 3ϕ circuit uses three diodes connected to a(n) _____ circuit with a neutral tap.

_____ 21. A photovoltaic cell is also known as a(n) _____.

_____ 22. _____ sensors detect the proximity of a magnetic field.

_____ 23. _____ light is light that is not visible to the human eye.

_____ 24. A(n) _____ pin is used for lead identification on a transistor.

_____ 25. Transistors were mainly developed to replace _____ switches.

_____ 26. _____ is the process of taking a small signal and increasing its size.

_____ 27. One advantage of a triac is that virtually no _____ is wasted by being converted to heat.

_____ 28. A(n) _____ is a very high-gain, directly-coupled amplifier.

_____ 29. A(n) _____ is an electronic circuit having two stable states designated set and reset.

_____ 30. A(n) _____ gate may be used in an elevator control circuit.

_____ 31. A(n) _____ diode is a light source for fiber-optic cables.

_____ 32. _____ consist of an IRED input stage and a silicon NPN phototransistor as the output stage.

_____ 33. A good diode has a(n) _____ drop across it when it is forward biased.

_____ 34. A(n) _____ should be tested using an oscilloscope if the device is suspected of being open.

_____ 35. Protons carry a(n) _____ charge.

_____ 36. _____ carry no electrical charge.

_____ 37. Electrons carry a(n) _____ charge.

_____ 38. _____ determine the conductive or insulative value of a given material.

_____ 39. The two types of material created by the addition of new atoms into a crystal are N-type and _____ material.

_____ 40. _____ flow is equal to and opposite of electron flow.

_____   **41.** _____ is the changing of AC into DC.

_____   **42.** A(n) _____ acts as a voltage regulator either by itself or in conjunction with other semiconductor devices.

_____   **43.** _____ direct current eliminates pulsations and provides direct current at a constant intensity.

_____   **44.** A(n) _____ rectifier circuit uses three rectifier diodes to convert AC to DC.

_____   **45.** The two types of transistors are the common/emitter and the _____.

_____   **46.** In any transistor circuit, the _____ junction must always be forward biased and the base/collector junction must always be reverse biased.

_____   **47.** The _____ is the critical factor in determining the amount of current flow in a transistor.

_____   **48.** A transistor switched ON is normally operating in the _____ region.

_____   **49.** When a transistor is switched OFF, it is operating in the _____ region.

_____   **50.** _____ amplifiers are two or more amplifiers used to obtain additional gain.

_____   **51.** A(n) _____ is a three-electrode AC semiconductor switch that conducts in both directions.

_____   **52.** A(n) _____ transistor is used primarily as a triggering device for SCRs and triacs.

_____   **53.** A(n) _____ is a bidirectional semiconductor that is used primarily as a triggering device.

_____   **54.** _____ are popular because they provide a complete circuit function in one semiconductor package.

_____   **55.** A(n) _____ amplifier is a very high-gain, directly-coupled amplifier that uses external feedback to control response characteristics.

_____   **56.** A(n) _____ gate is a device with an output that is high only when both of its inputs are high.

_____   **57.** A(n) _____ gate is a device with an output that is high if either or both inputs are high.

_____   **58.** A(n) _____ gate is the same as an inverted OR function.

_____   **59.** A(n) _____ gate is an inverted AND function.

_____   **60.** A(n) _____ is an integrated circuit designed to output timing pulses for control of certain types of circuits.

_____   **61.** _____ are the smallest building blocks of matter.

   A. Atoms   C. Electrons
   B. Molecules   D. neither A, B, nor C

_____   **62.** _____ current is current passed when a diode breaks down.

   A. Breakover   C. Blocking
   B. Avalanche   D. Holding

_____ 63. A(n) _____ is a circuit containing a diode which permits only the positive half-cycles of the AC sine wave to pass.

_____ 64. A(n) _____ is the area on a semiconductor material between the P-type and N-type material.

_____ 65. A(n) _____ is a device which conducts current when energized by light.

_____ 66. A(n) _____ is a diode which is switched ON and OFF by light.

_____ 67. A(n) _____ is a three-terminal device that controls current through the device depending on the amount of voltage applied to the base.

_____ 68. A(n) _____ is a solid-state rectifier with the ability to rapidly switch heavy currents.

_____ 69. _____ voltage is the voltage required to switch an SCR into a conductive state.

_____ 70. _____ is a technology that uses a thin flexible glass or plastic fiber to transmit light.

Solid-State Component Identification

_____ 1. Thermistor

_____ 2. Photoconductive cell

_____ 3. Photovoltaic cell

_____ 4. Photoconductive diode

_____ 5. Hall effect sensor

_____ 6. Solid-state pressure sensor

_____ 7. Laser diode

_____ 8. Pin photodiode

_____ 9. Phototransistor

_____ 10. Light-activated SCR

_____ 11. Phototriac

_____ 12. Optocoupler

A. Responds to magnetic influence

B. Converts solar energy to electrical energy

C. A thermally-sensitive resistor

D. Changes resistance with pressure

E. A light-sensitive resistor

F. A light-sensitive diode

G. Combines effect of photodiode and transistor

H. Light sensitive gate and bidirectional

I. Provides electrical isolation between circuits

J. LASCR

K. Produces coherent light

L. Light radiation disturbs the PN junction

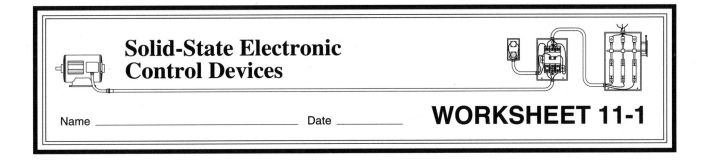

Solid-State Electronic Control Devices

Name _____ Date _____

WORKSHEET 11-1

PC Boards

_____ 1. Insulated board

_____ 2. Traces

_____ 3. Terminal contacts

_____ 4. Pads

_____ 5. Bus

_____ 6. Edge card connector

_____ 7. Components

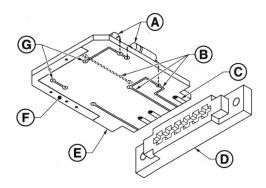

Solid-State Components

_____ 1. Resistors

_____ 2. Wafer switch

_____ 3. Power transformer

_____ 4. Transistor

_____ 5. SCR

_____ 6. Mini-DIP IC

_____ 7. TO-5 IC

_____ 8. Trimmer resistor

_____ 9. Capacitor

_____ 10. Dual-in-line IC

_____ 11. Large scale IC

_____ 12. Heat sink

_____ 13. Dip switch

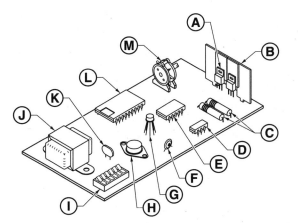

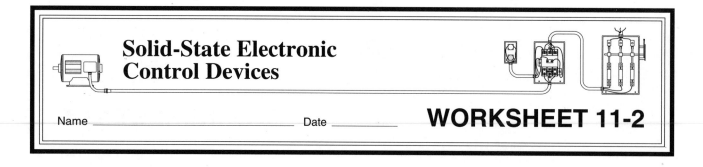

Solid-State Electronic Control Devices

Name _____ Date _____

WORKSHEET 11-2

Diode Operation

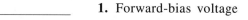

_____ **1.** Forward-bias voltage

_____ **2.** Reverse breakdown

_____ **3.** Reverse-bias voltage

_____ **4.** Reverse current

_____ **5.** Forward operating current

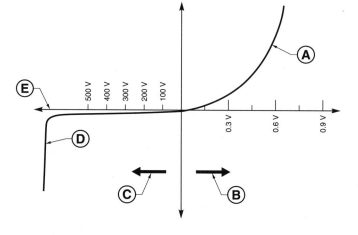

Zener Diode Operation

_____ **1.** Zener breakdown

_____ **2.** Forward breakover voltage

_____ **3.** Forward current

_____ **4.** Standard diode operating range

_____ **5.** Reverse-bias voltage

_____ **6.** Forward-bias voltage

_____ **7.** Zener diode operating range

_____ **8.** Reverse current

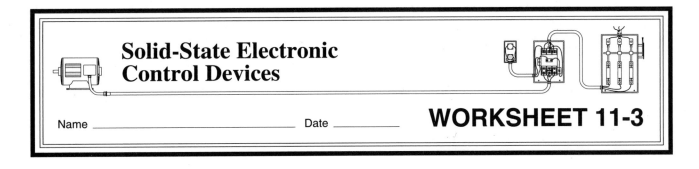

Solid-State Electronic Control Devices

Name _____ Date _____ **WORKSHEET 11-3**

SCR Operation

_____   1. Forward breakover voltage

_____   2. Holding current

_____   3. Reverse breakdown voltage

_____   4. Avalanche current

_____   5. Reverse current

_____   6. Forward blocking current

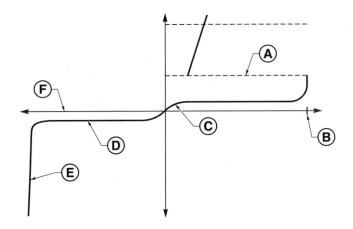

Amplifiers

1. Connect the load to the common/emitter amplifier.

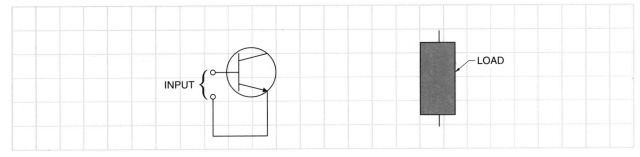

2. Connect the load to the common/base amplifier.

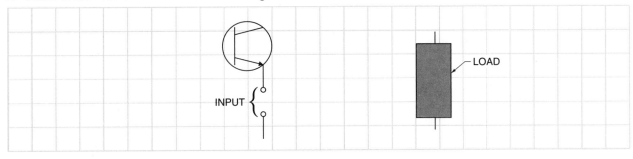

Solid-State Electronic Control Devices

Name _____ Date _____

Solid-State Devices

1. Complete the schematic for the 555 timer and label each part.

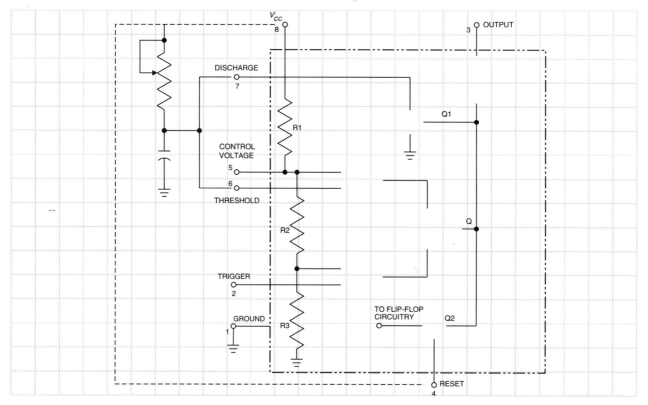

2. Indicate the direction of current flow through the full-wave bridge rectifier. A is positive and B is negative with respect to A. D1 and D2 are forward biased and conduct. D3 and D4 are reversed biased and do not conduct.

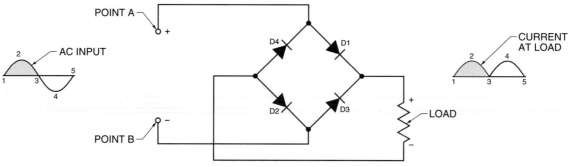

Solid-State Electronic Control Devices

Name _____ Date _____

WORKSHEET 11-5

Solid-State Symbol Identification

_____ 1. SCR

_____ 2. Thermistor

_____ 3. Diode

_____ 4. Optocoupler

_____ 5. Triac

_____ 6. LED

_____ 7. Unijunction transistor

_____ 8. Diac

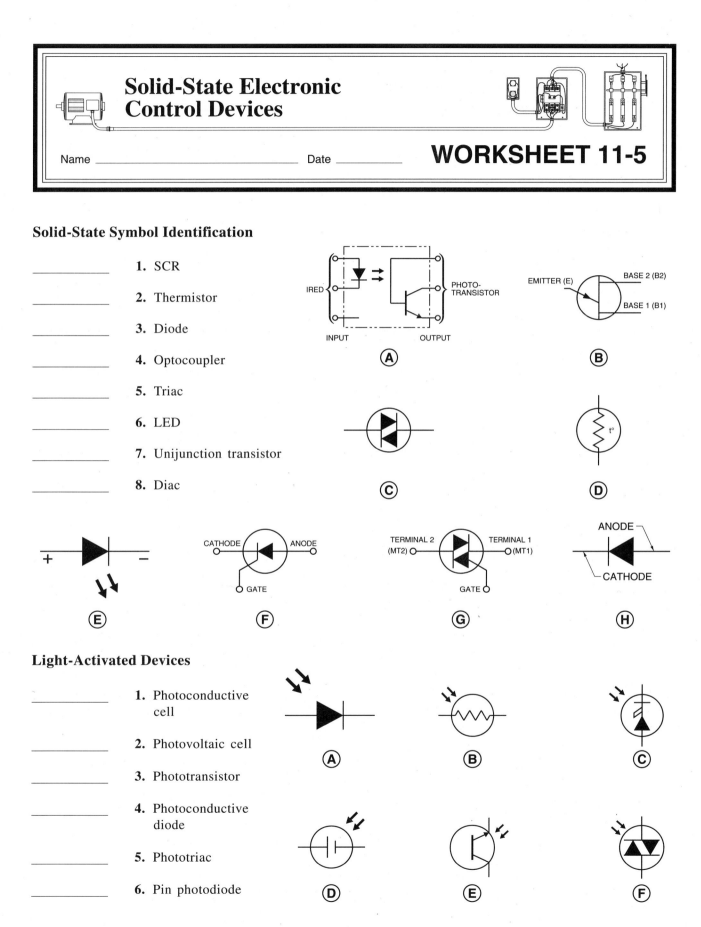

Light-Activated Devices

_____ 1. Photoconductive cell

_____ 2. Photovoltaic cell

_____ 3. Phototransistor

_____ 4. Photoconductive diode

_____ 5. Phototriac

_____ 6. Pin photodiode

Integrated Circuits

_____ **1.** Dual-in-line

_____ **2.** Mini-DIP

_____ **3.** TO-5

_____ **4.** MOS/LSI

_____ **5.** Flat pack

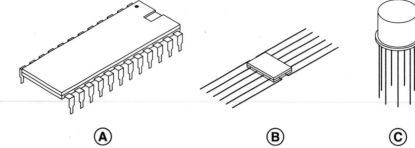

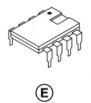

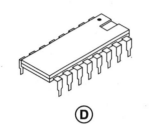

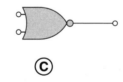

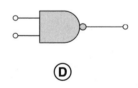

Logic Gates

_____ **1.** NAND

_____ **2.** NOR

_____ **3.** OR

_____ **4.** AND

Transistors

_____ **1.** Base

_____ **2.** Collector

_____ **3.** Emitter

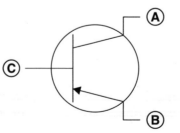

Electromechanical and Solid-State Relays

Name _____   Date _____

TECH-CHEK 12

Electrical Motor Controls

_____   **1.** _____ are the number of completely isolated circuits that can pass through a switch at one time.

A. Throws
B. Poles
C. Makes
D. Breaks

_____   **2.** _____ are the number of closed contact positions per pole that are available on a switch.

A. Throws
B. Poles
C. Makes
D. Breaks

_____   **3.** _____ are the number of separate contacts a switch uses to open or close each individual circuit.

A. Throws
B. Poles
C. Makes
D. Breaks

_____   **4.** A(n) _____ relay is used to ramp up the voltage applied to a load.

A. reed
B. general purpose
C. instant ON
D. analog switching

_____   **5.** A(n) _____ relay is used to turn ON a load when the load voltage crosses near or at the zero point.

A. reed
B. instant ON
C. machine control
D. zero switching

_____   **6.** A(n) _____ relay is normally a plug-in relay and does not have convertible or replaceable contacts.

A. reed
B. general purpose
C. instant ON
D. peak switching

_____   **7.** A(n) _____ relay normally includes replaceable, convertible, or interchangeable contacts.

A. analog switching
B. machine control
C. general purpose
D. zero switching

_____   **8.** A(n) _____ relay is activated by the presence of a magnetic field.

A. reed
B. general purpose
C. instant ON
D. peak switching

_____   **9.** A(n) _____ relay is an SSR that allows the load to be turned ON at any point on the AC sine wave.

A. reed
B. peak switching
C. instant ON
D. zero switching

Reed Relay Actuation

_____ 1. Parallel motion

_____ 2. Perpendicular motion

_____ 3. Shielding

_____ 4. Pivoted motion

_____ 5. Rotary motion

_____ 6. Front-to-back motion

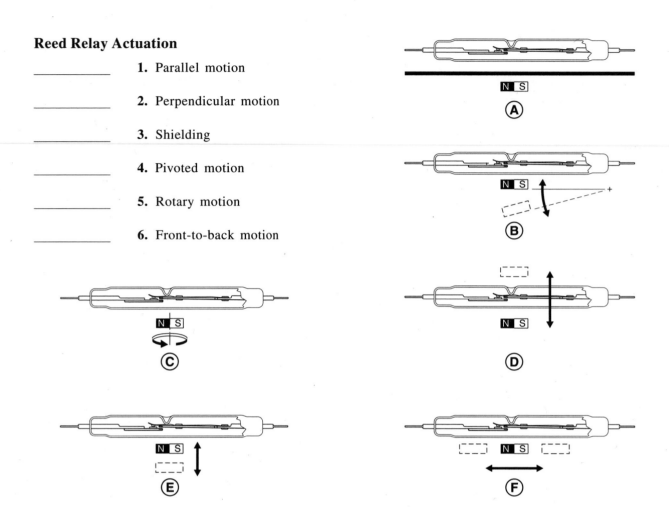

EMR and SSR Comparison

_____ 1. Coil voltage

_____ 2. Coil current

_____ 3. Pull-in time

_____ 4. Contact voltage rating

_____ 5. Contact current rating

_____ 6. Contact voltage drop

_____ 7. Insulation resistance

A. Switch-OFF resistance

B. Switch-ON voltage drop

C. Control voltage

D. Load current

E. Load voltage

F. Control current

G. Turn-ON time

Electromechanical and Solid-State Relays

Name _____   Date _____

WORKSHEET 12-1

Electromechanical Relay Contact Addition

Complete the line diagrams according to the circuit information. Use standard lettering, numbering, and coding information.

1. Design a control circuit so a 24 V pushbutton (ON/OFF with memory) operates an electromechanical relay to control a 230 V solenoid.

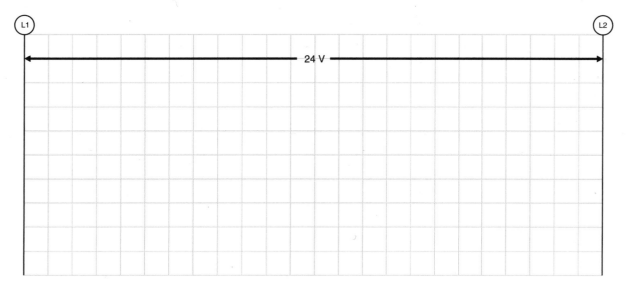

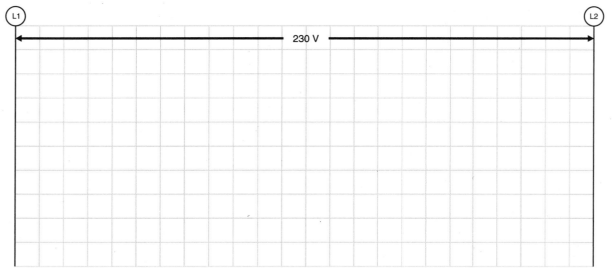

2. Design a control circuit so a magnetic motor starter is activated by a standard start/stop pushbutton station with memory. Add a control relay with three auxiliary contacts to extend the number of contacts available with the motor starter. One set of NC contacts is to control a green light, the other set of NC contacts is to control a solenoid, and the NO contacts are to control a red light. Provide overload protection for the motor.

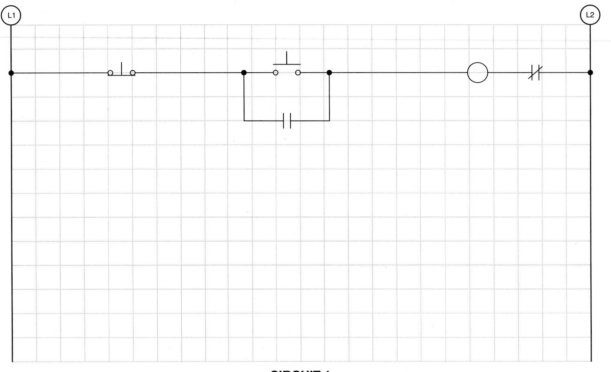

CIRCUIT 1

3. Redraw Circuit 1, adding a 24 VAC relay coil and motor starter coil to control the motor and a 115 VAC green light, red light, and solenoid into the circuit.

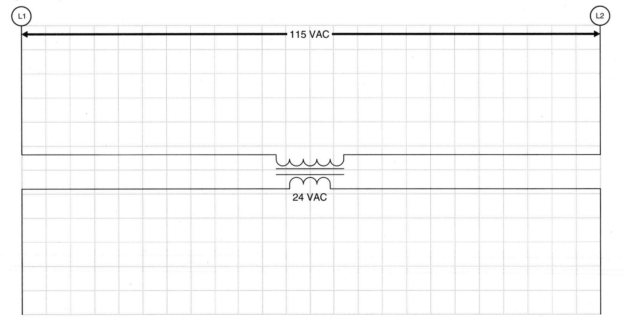

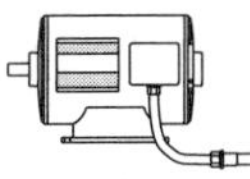

Electromechanical and
Solid-State Relays

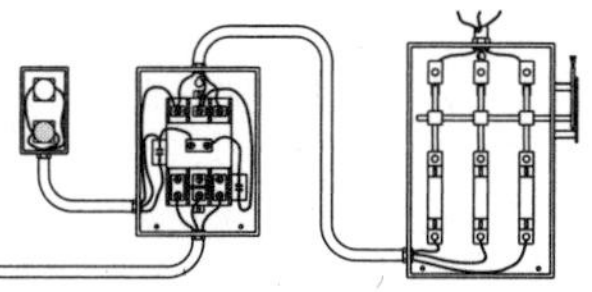

Name _____ Date _____

WORKSHEET 12-2

Pole/Throw/Break Identification

Determine the correct number of poles, throws, and breaks for each contact arrangement.

_____ **1.** Poles

_____ **2.** Throws

_____ **3.** Breaks

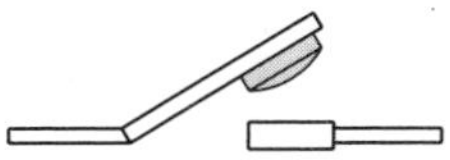

_____ **4.** Poles

_____ **5.** Throws

_____ **6.** Breaks

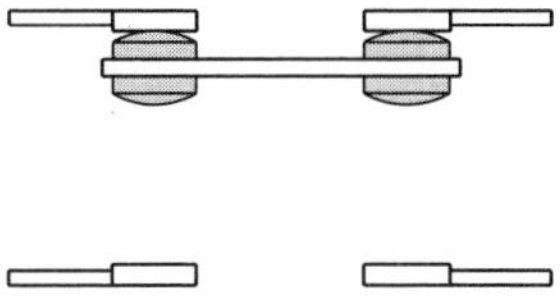

_____ **7.** Poles

_____ **8.** Throws

_____ **9.** Breaks

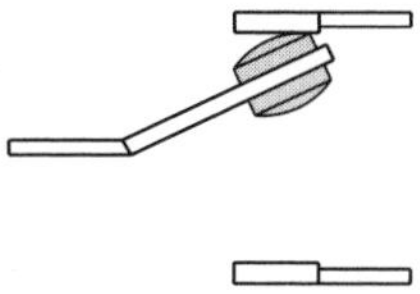

_____ **10.** Poles

_____ **11.** Throws

_____ **12.** Breaks

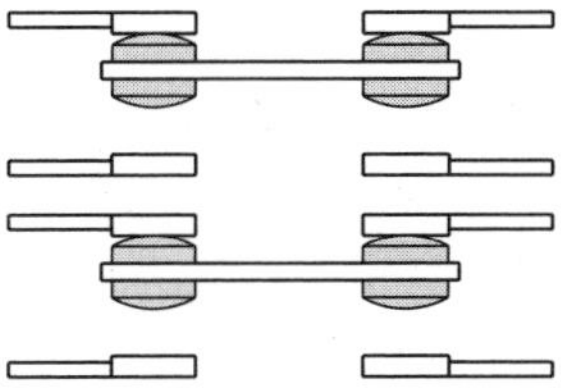

_____ **13.** Poles

_____ **14.** Throws

_____ **15.** Breaks

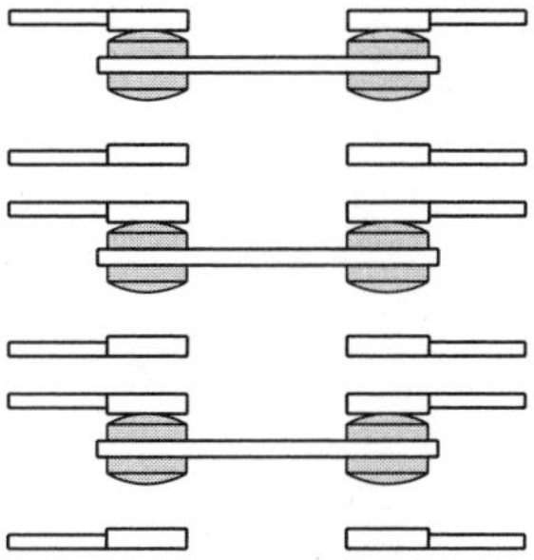

_____ **16.** Poles

_____ **17.** Throws

_____ **18.** Breaks

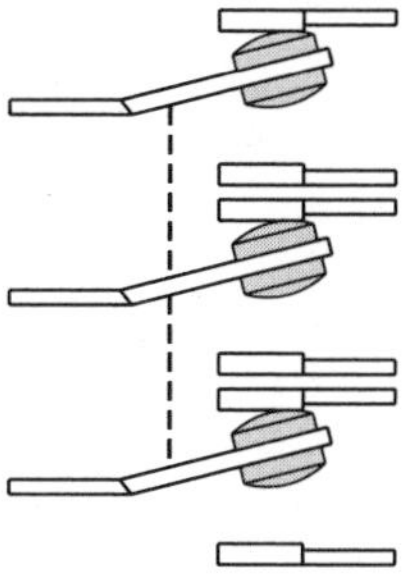

_____ **19.** Poles

_____ **20.** Throws

_____ **21.** Breaks

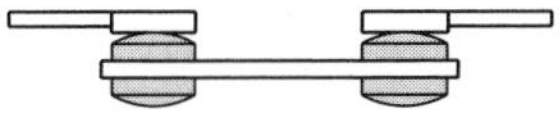

_____ **22.** Poles

_____ **23.** Throws

_____ **24.** Breaks

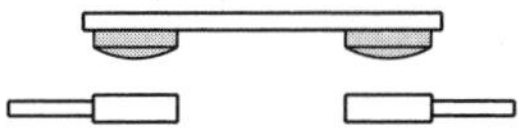

Electromechanical and Solid-State Relays

Name _____ Date _____

WORKSHEET 12-3

Control Circuit Solid-State Relay Use

Complete the control diagrams according to the circuit information. Use standard lettering, numbering, and coding information.

1. Complete the control diagram so it contains a start pushbutton connected for memory to start the motor and a stop pushbutton to stop the motor. Include overload contacts to stop the motor if an overload occurs. Use an SCR for solid-state memory. Include a current-limiting resistor in the gate circuit.

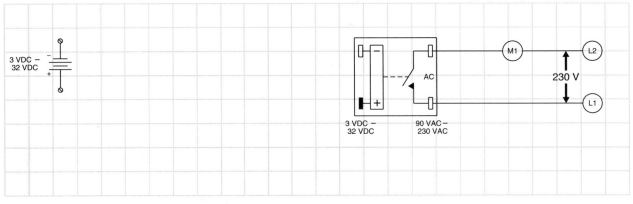

CIRCUIT 1

2. Complete the wiring diagram showing the actual wire placement for Circuit 1.

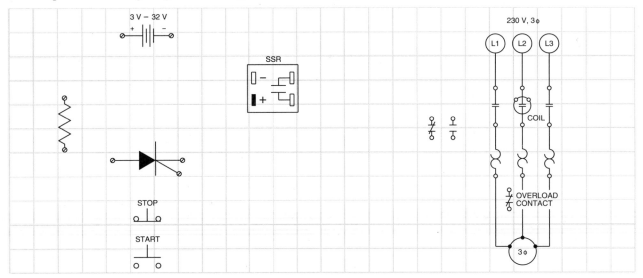

3. Redraw Circuit 1, adding an ON-delay timer to automatically turn OFF the motor after 30 min. The timer has a coil voltage rated at the same voltage level as the magnetic motor starter coil.

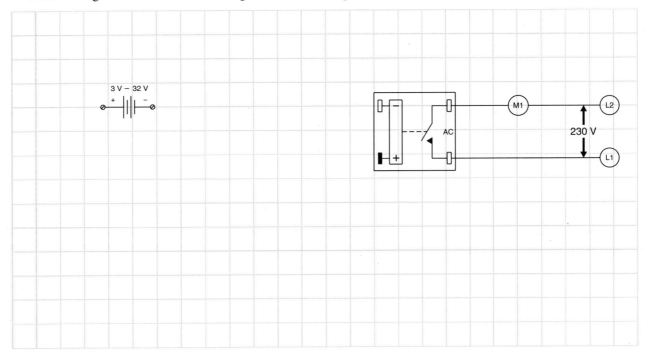

4. Redraw Circuit 1, adding an OFF-delay timer to keep the motor running for 30 min after the stop pushbutton is pressed. The timer has a coil voltage rated at the same voltage level as the magnetic motor starter coil.

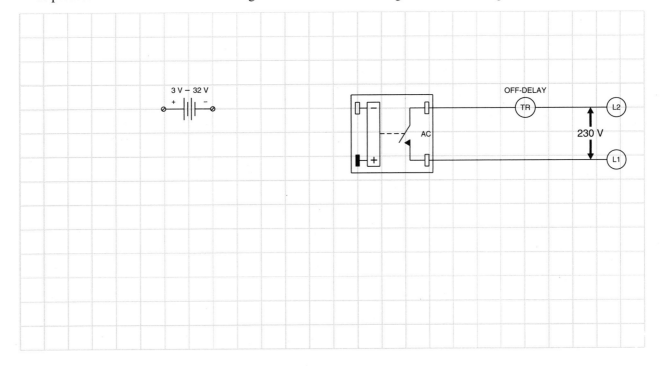

Electromechanical and Solid-State Relays

Name _____ Date _____

WORKSHEET 12-4

Forward and Reversing Solid-State Relay Use

Complete the control diagram according to the circuit information. Use standard lettering, numbering, and coding information.

1. Design a control circuit using two solid-state relays to forward and reverse a motor. Include a forward/reverse/stop pushbutton and pushbutton interlocking. Use SCRs to add memory in the forward and reverse direction. Include a current limiting resistor in each gate circuit.

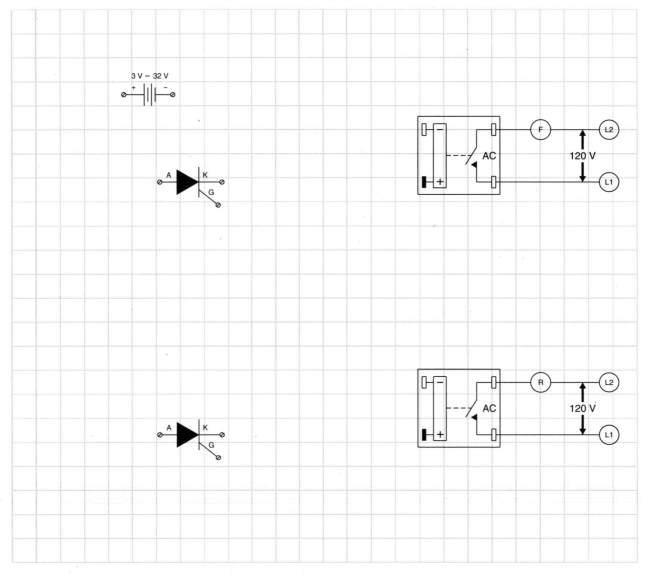

Photoelectric and Proximity Controls

Name _____ Date _____

TECH-CHEK 13

Electrical Motor Controls

_____ 1. The control operating mode is _____ if the output is energized when the light beam is not blocked and the photosensor is illuminated.

    A. light operated   C. ON
    B. dark operated   D. OFF

_____ 2. The control operating mode is _____ if the output is energized when the light beam is blocked and the photosensor is not illuminated.

    A. light operated   C. ON
    B. dark operated   D. OFF

_____ 3. The operating point at which the level of light intensity triggers the output is determined by the _____ adjustment.

    A. sensitivity   C. voltage
    B. differential   D. current

_____ 4. _____ scan is a scanning technique which uses a special filter that emits a beam of light so it is projected in one plane only.

    A. Direct   C. Polarized
    B. Retroreflective   D. Diffuse

_____ 5. _____ scan is a scanning technique normally used in high-vibration applications.

    A. Direct   C. Specular
    B. Retroreflective   D. Convergent beam

_____ 6. _____ scan is a scanning technique which should generally be the first choice for scanning targets that block most of the light beam.

    A. Direct   C. Specular
    B. Polarized   D. Convergent beam

_____ 7. _____ scan is a scanning technique which is generally used in color-mark detection.

    A. Retroreflective   C. Specular
    B. Polarized   D. Diffuse

_____ 8. _____ scan is a scanning technique which places the transmitter and receiver at equal angles from a highly-reflective surface.

    A. Direct   C. Specular
    B. Retroreflective   D. Diffuse

_____ 9. _____ scan is a scanning technique which focuses the light beam to a fixed focal point in front of the photoreceiver.

    A. Polarized   C. Diffuse
    B. Specular   D. Convergent beam

_____ **10.** A(n) _____ sensor is a proximity switch that detects a magnetic field.

    A. inductive   C. Hall effect
    B. capacitive   D. pendulum

_____ **11.** A(n) _____ sensor is a proximity switch that detects conductive substances only.

    A. inductive   C. Hall effect
    B. capacitive   D. pendulum

_____ **12.** A(n) _____ sensor is a proximity switch that detects either conductive or non-conductive substances.

    A. inductive   C. Hall effect
    B. capacitive   D. pendulum

Scanning Methods

_____ **1.** Direct scan

_____ **2.** Retroreflective scan

_____ **3.** Polarized scan

_____ **4.** Specular scan

_____ **5.** Diffuse scan

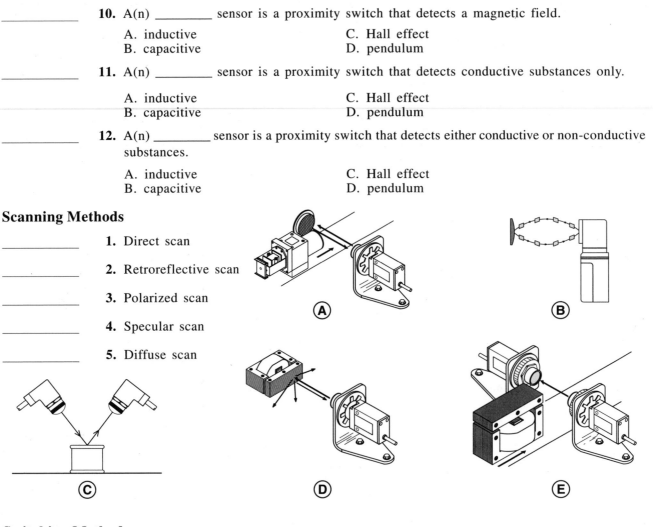

Switching Methods

_____ **1.** Two-wire AC switching

_____ **2.** NPN transistor switching

_____ **3.** PNP transistor switching

_____ **4.** Photoelectric relay module switching

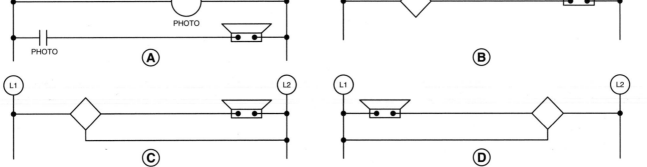

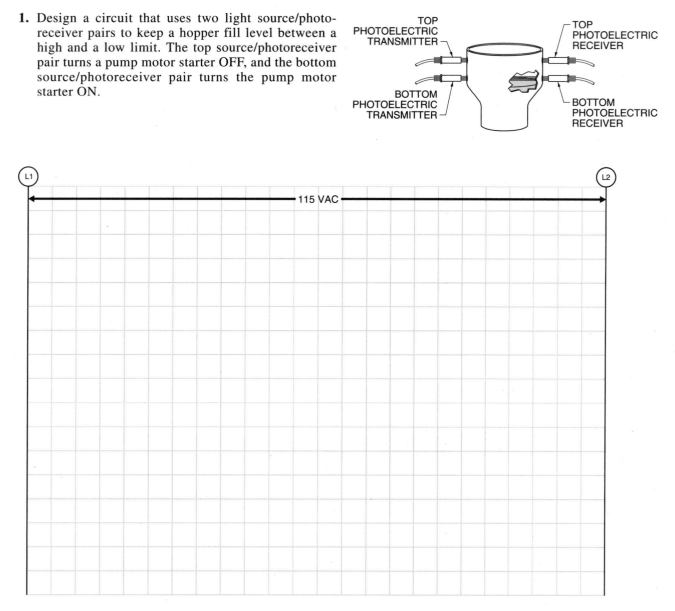

Photoelectric and Proximity Controls

Name _____ Date _____

WORKSHEET 13-1

Level Control

Use Data Sheet I to complete the line diagram. Use standard lettering, numbering, and coding information. Mark all manufacturer's numbers (1 – 11) on the line diagram.

1. Design a circuit that uses two light source/photo-receiver pairs to keep a hopper fill level between a high and a low limit. The top source/photoreceiver pair turns a pump motor starter OFF, and the bottom source/photoreceiver pair turns the pump motor starter ON.

TOP PHOTOELECTRIC TRANSMITTER

TOP PHOTOELECTRIC RECEIVER

BOTTOM PHOTOELECTRIC TRANSMITTER

BOTTOM PHOTOELECTRIC RECEIVER

L1

L2

115 VAC

Photoelectric and Proximity Controls

Name _____ Date _____

WORKSHEET 13-2

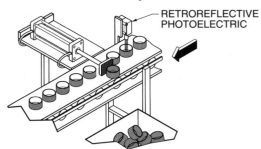

Part Checking

Use Data Sheet I to complete the line diagram. Use standard lettering, numbering, and coding information. Mark all manufacturer's numbers (1 – 11) on the line diagram. Assume there is no space between the cans.

1. Design a circuit so that dark caps are checked for white liners by a photoelectric scanner. The scanner activates a solenoid valve, which controls a cylinder that rejects caps that lack liners. The solenoid is activated 5 sec after the scanner sees a cap without a liner. The white liners act to stop the reflection from the tin bottom. The tin bottom acts as a reflector.

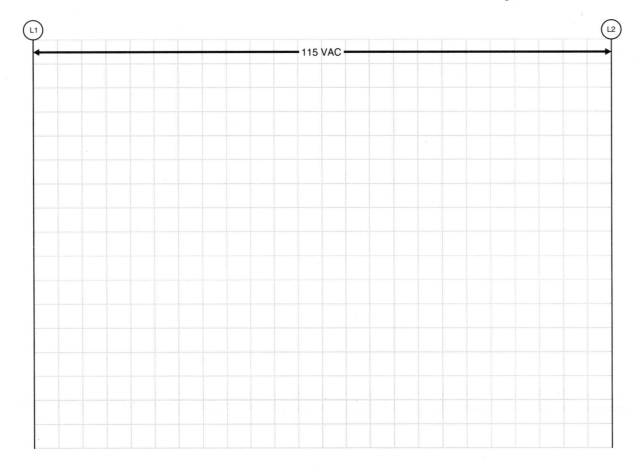

RETROREFLECTIVE PHOTOELECTRIC

L1   L2

← 115 VAC →

Photoelectric and Proximity Controls

Name _____ Date _____

WORKSHEET 13-3

Carton Filling

Use Data Sheet I to complete the line diagram. Use standard lettering, numbering, and coding information. Mark all manufacturer's numbers (1 – 11) on the line diagram. Label each timer as to its time setting.

1. Design a photoelectric control that stops a conveyor motor and fills a carton. The photoelectric control does not stop the motor until 2 sec (timer one) after it sees the carton. The fill process is controlled by a timer that controls a fill solenoid for a preset time duration. It takes 15 sec (timer two) to fill the carton. After the carton is filled, the fill solenoid is turned OFF, the conveyor motor is turned ON, and the timer is reset.

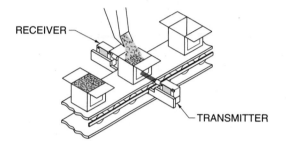

RECEIVER

TRANSMITTER

L1 ──── 115 VAC ──── L2

Photoelectric and Proximity Controls

Name _____ Date _____

WORKSHEET 13-4

Carton Gluing

Use Data Sheet I to complete the line diagram. Use standard lettering, numbering, and coding information. Mark all manufacturer's numbers (1 – 11) on the line diagram. Label each timer as to its time setting.

1. Design a photoelectric control circuit that turns ON a glue nozzle when a carton passes. The glue nozzle is controlled by a solenoid valve. The conveyor motor is ON at all times. Do not show the conveyor circuit in the line diagram. The glue operation does not start until 2 sec (timer one) after the photoelectric control sees the carton. The glue operation stops after 5 sec (timer two), even though the carton takes 9 sec to pass the photoelectric control. This prevents glue from being sprayed on the conveyor belt.

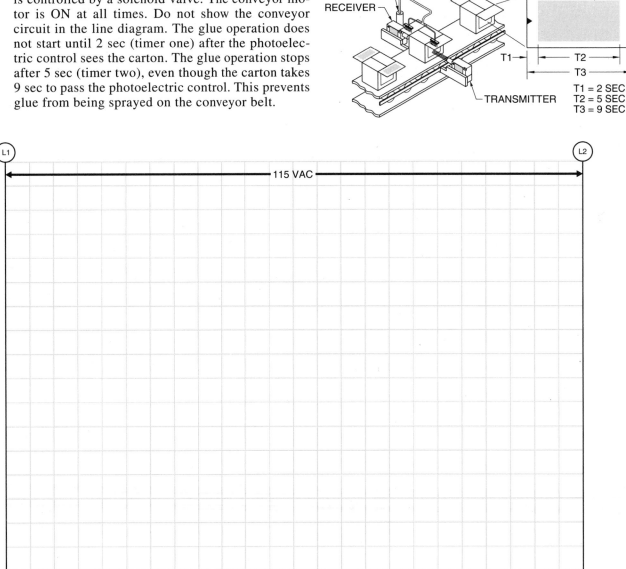

GLUE SOLENOID · RECEIVER · FLAP · DESIRED GLUE PATTERN ON FLAP · TRANSMITTER · T1 · T2 · T3

T1 = 2 SEC
T2 = 5 SEC
T3 = 9 SEC

L1 ← 115 VAC → L2

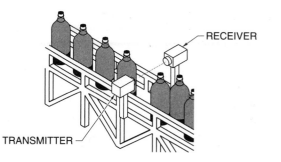

Photoelectric and Proximity Controls

Name _____ Date _____

WORKSHEET 13-5

Conveyor Jam Control

Use Data Sheet I to complete the line diagram. Use standard lettering, numbering, and coding information. Mark all manufacturer's numbers (1 – 11) on the line diagram.

1. Design a photoelectric control circuit that sounds an alarm if there is a jam or if there are no products moving along a conveyor system. Normally-spaced products move past the photoelectric control (set for dark operated) every 1 sec to 3 sec. The circuit sounds the alarm if no product has moved past the control in 9 sec (use an ON-delay timer set for 9 sec) or if a product stays in front of the control for more than 6 sec (use an ON-delay timer set for 6 sec). Mark the 6 sec timer TR1 and the 9 sec timer TR2.

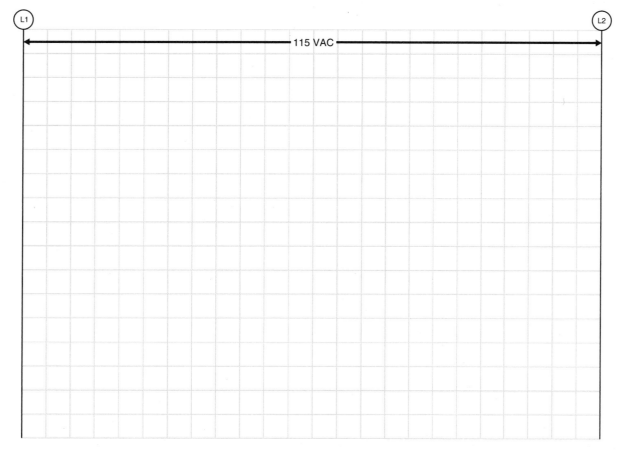

L1   115 VAC   L2

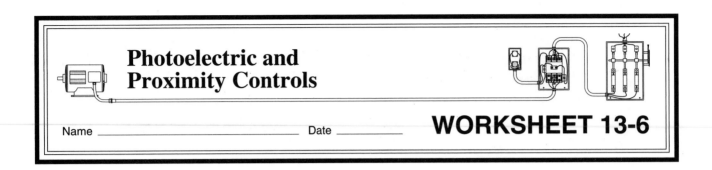

Photoelectric and Proximity Controls

Name _____ Date _____

WORKSHEET 13-6

Paper Roll Monitoring

Use Data Sheet I to complete the line diagram. Use standard lettering, numbering, and coding information. Mark all manufacturer's numbers (1 – 11) on the line diagram.

1. Design a circuit in which the position of the photo-electric control (set for light operated) monitors the diameter of a roll of paper. The control should turn OFF a drive motor and sound an alarm when the roll of paper is almost empty. Include a standard start/stop pushbutton station to start the drive motor and stop it manually if required.

TRANSMITTER   RECEIVER

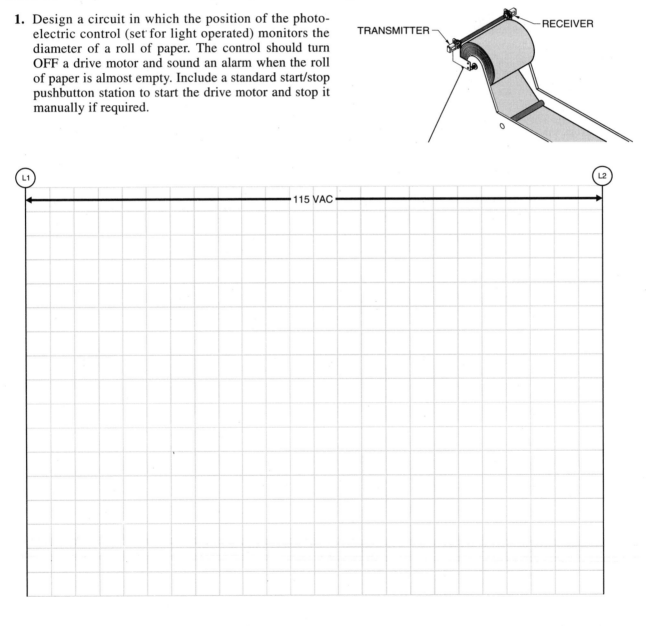

L1   115 VAC   L2

Photoelectric and Proximity Controls

Name _____ Date _____

WORKSHEET 13-7

Steel Cutting Operation

Use Data Sheet J to complete the line diagram. Use standard lettering, numbering, and coding information. Mark all manufacturer's numbers (1 – 11) on the line diagram.

1. Design a circuit with a proximity switch that is used to automatically stop a motor from supplying steel to a cutting press. A start pushbutton is used to start the steel infeed, and a proximity switch or stop pushbutton is used to stop the steel infeed.

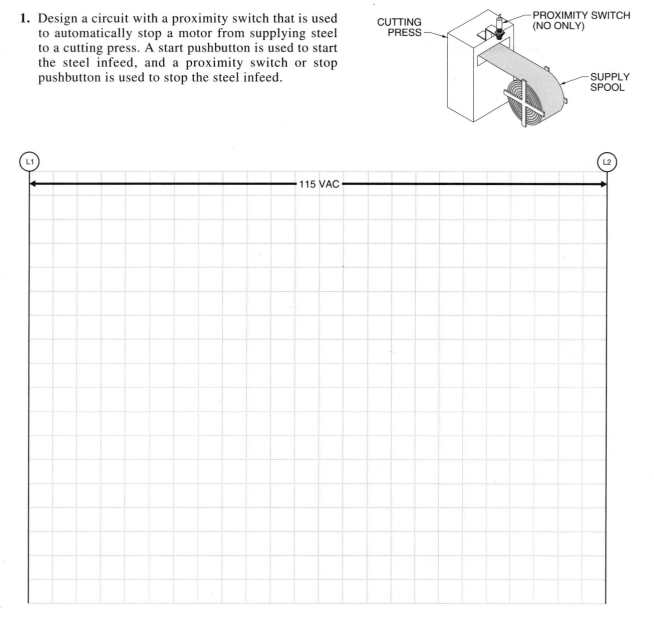

CUTTING PRESS

PROXIMITY SWITCH (NO ONLY)

SUPPLY SPOOL

L1

L2

115 VAC

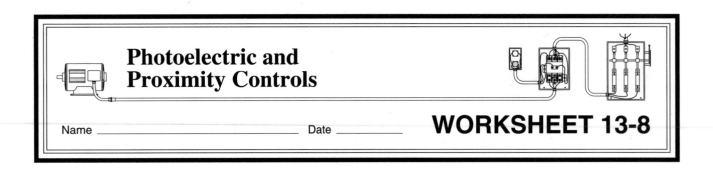

Photoelectric and Proximity Controls

Name _____ Date _____

WORKSHEET 13-8

Bottle Cap Control

Use Data Sheets I and J to complete the line diagram. Use standard lettering, numbering, and coding information. Mark all manufacturer's numbers (1 – 11) on the line diagram.

1. Design a circuit in which an alarm sounds if a bottle cap is missing. The photoelectric control (set for dark operated) is used to detect if a bottle is present. The proximity control detects if there is a cap on the bottle. Show both the photoelectric and proximity controls in the circuit.

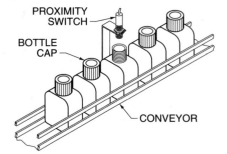

PROXIMITY SWITCH

BOTTLE CAP

CONVEYOR

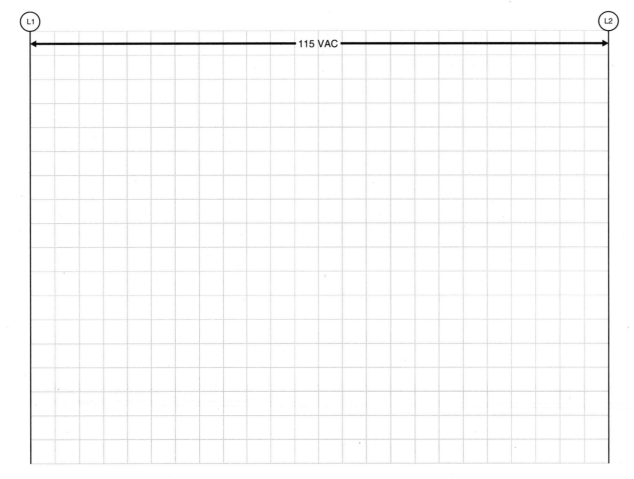

L1   115 VAC   L2

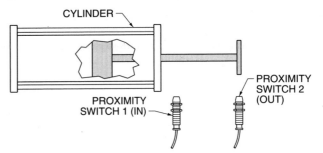

Photoelectric and Proximity Controls

Name _____ Date _____

WORKSHEET 13-9

Piston Control

Use Data Sheet J to complete the line diagram. Use standard lettering, numbering, and coding information. Mark all manufacturer's numbers (1 – 11) on the line diagram.

1. Design a circuit in which two proximity switches are used to cycle the piston in a cylinder back and forth automatically. A standard start/stop pushbutton station with memory is used to start and stop the automatic cycling. The piston is controlled by a double solenoid fluid power valve. One solenoid controls the piston out function and the other controls the piston in function. Power must be maintained on the solenoids to keep the piston moving.

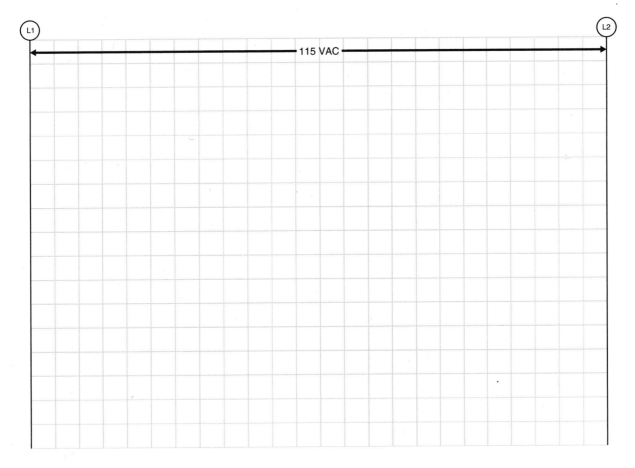

Photoelectric and Proximity Controls

Name _____ Date _____ **WORKSHEET 13-10**

Proximity Switch Piston Control

Use Data Sheet K to complete the line diagram. Use standard lettering, numbering, and coding information. Mark all manufacturer's numbers (1 – 11) on the line diagram.

1. Design a circuit in which two proximity switches are used to cycle the piston in a cylinder back and forth automatically. A standard start/stop pushbutton station with memory is used to start and stop the automatic cycling. The piston is controlled by a double solenoid fluid power valve. One solenoid controls the piston out function and the other controls the piston in function. Power must be maintained on the solenoids to keep the piston moving.

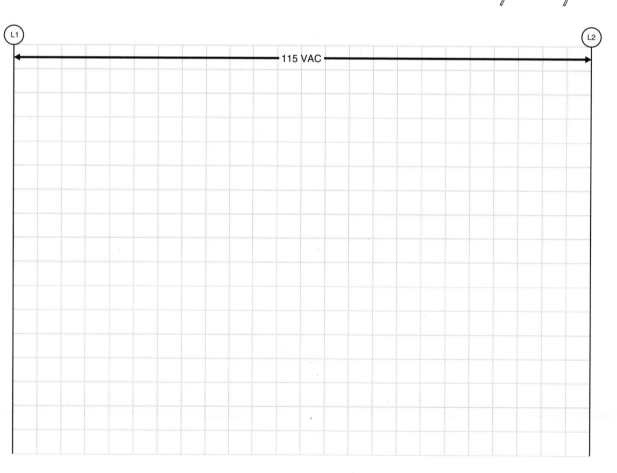

Programmable Controllers

Name _____ Date _____

TECH-CHEK 14

Electrical Motor Controls

_____ 1. _____ manufacturing produces such goods as automobiles and refrigerators.

    A. Movable   C. Process
    B. Multiple   D. Discrete parts

_____ 2. _____ manufacturing produces such goods as food and gas.

    A. Movable   C. Process
    B. Multiple   D. Discrete parts

_____ 3. The _____ section of a PLC organizes all system control activities.

    A. power supply   C. processor
    B. input/output   D. programming

_____ 4. The _____ section of a PLC allows inputs into the PLC through a keyboard.

    A. power supply   C. processor
    B. input/output   D. programming

_____ 5. The _____ section of a PLC provides the voltage required for internal operation and charging the internal battery.

    A. power supply   C. processor
    B. input/output   D. programming

_____ 6. The _____ section of a PLC functions as the eyes, ears, and hands of the PLC.

    A. power supply   C. processor
    B. input/output   D. programming

_____ 7. The _____ mode is used to execute the program in a PLC.

    A. program   C. test
    B. run   D. fast

_____ 8. The _____ mode is used when developing the logic of a circuit.

    A. program   C. test
    B. run   D. fast

_____ 9. The _____ mode is used when forcing inputs and outputs.

    A. program   C. test
    B. run   D. fast

_____ 10. A _____ system has multiple transmitters and receivers connected to a single-wire pair.

    A. hard-wired   C. hand-operated
    B. self-diagnosis   D. multiplexing

PLC Logic Functions

_____ **1.** AND logic

_____ **2.** OR logic

_____ **3.** NOT logic

_____ **4.** NAND logic

_____ **5.** NOR logic

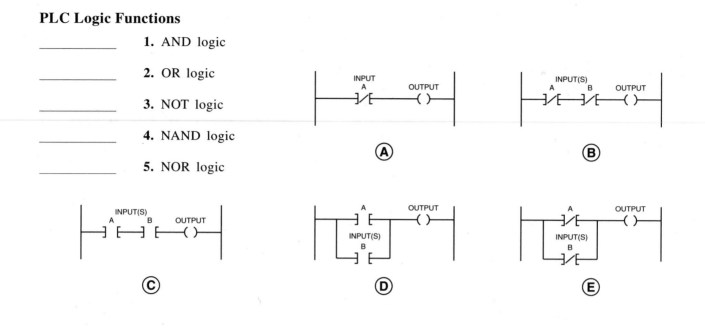

PLC Programming Diagrams

1. Redraw the standard line diagram as a basic PLC programming diagram.

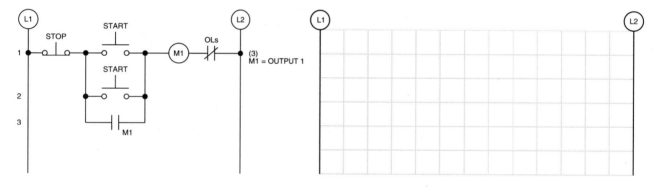

2. Redraw the basic PLC programming diagram as a standard line diagram.

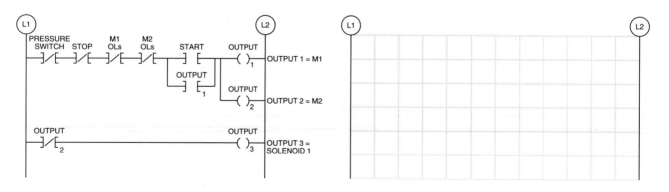

Programmable Controllers

Name _____ Date _____

WORKSHEET 14-1

Line Diagram to PLC Diagram Conversion

Use the Device Equivalents table to complete the PLC programming diagram.

1. Draw the basic PLC programming diagram of the standard line diagram.

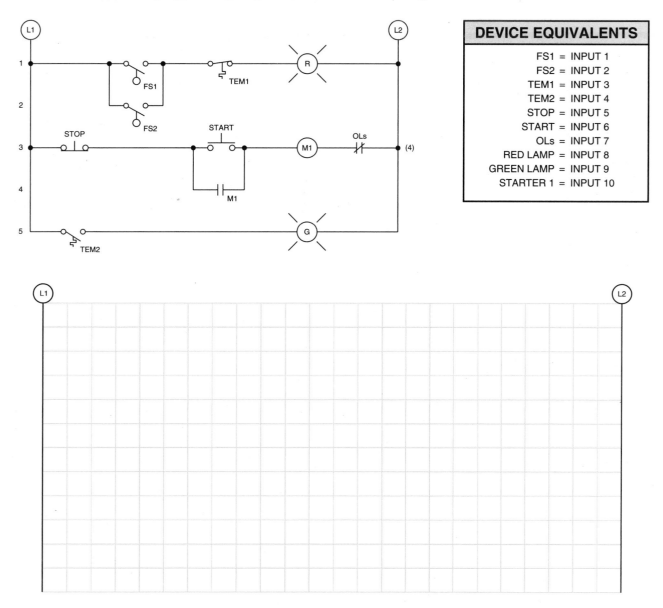

DEVICE EQUIVALENTS

FS1	= INPUT 1
FS2	= INPUT 2
TEM1	= INPUT 3
TEM2	= INPUT 4
STOP	= INPUT 5
START	= INPUT 6
OLs	= INPUT 7
RED LAMP	= INPUT 8
GREEN LAMP	= INPUT 9
STARTER 1	= INPUT 10

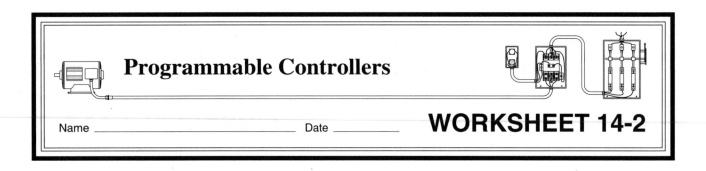

Programmable Controllers

Name _____ Date _____

WORKSHEET 14-2

Selector Switch Line Diagram to PLC Diagram Conversion

Use the Device Equivalents table to complete the PLC programming diagram.

1. Draw the basic PLC programming diagram of the standard line diagram.

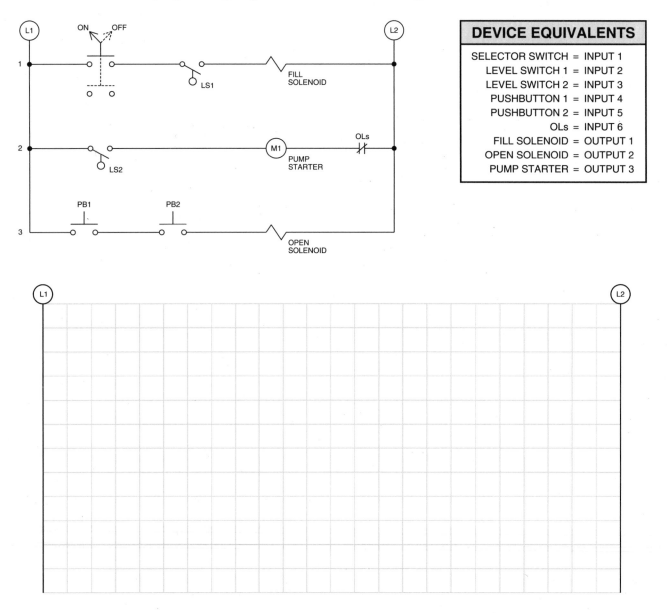

DEVICE EQUIVALENTS
SELECTOR SWITCH = INPUT 1
LEVEL SWITCH 1 = INPUT 2
LEVEL SWITCH 2 = INPUT 3
PUSHBUTTON 1 = INPUT 4
PUSHBUTTON 2 = INPUT 5
OLs = INPUT 6
FILL SOLENOID = OUTPUT 1
OPEN SOLENOID = OUTPUT 2
PUMP STARTER = OUTPUT 3

Programmable Controllers

Name _____ Date _____

WORKSHEET 14-3

Two-Speed Line Diagram to PLC Diagram Conversion

Use the Device Equivalents table to complete the PLC programming diagram.

1. Draw the basic PLC programming diagram of the standard line diagram.

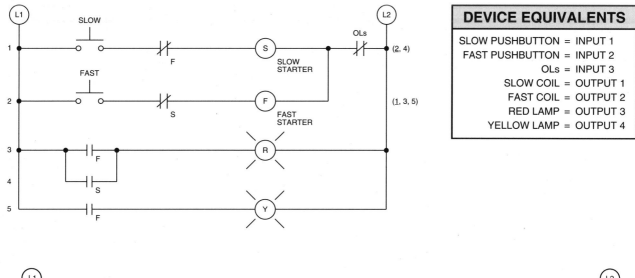

DEVICE EQUIVALENTS	
SLOW PUSHBUTTON =	INPUT 1
FAST PUSHBUTTON =	INPUT 2
OLs =	INPUT 3
SLOW COIL =	OUTPUT 1
FAST COIL =	OUTPUT 2
RED LAMP =	OUTPUT 3
YELLOW LAMP =	OUTPUT 4

Programmable Controllers

Name _____ Date _____

WORKSHEET 14-4

Forward and Reversing Line Diagram to PLC Diagram Conversion

Use the Device Equivalents table to complete the PLC programming diagram.

1. Draw the basic PLC programming diagram of the standard line diagram.

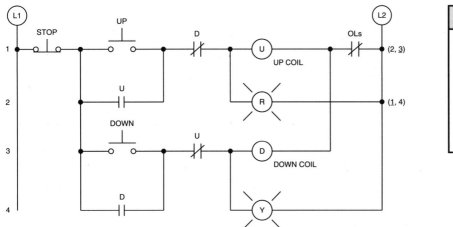

DEVICE EQUIVALENTS

STOP PUSHBUTTON = INPUT 1
UP PUSHBUTTON = INPUT 2
DOWN PUSHBUTTON = INPUT 3
OLs = INPUT 4
UP COIL = OUTPUT 1
DOWN COIL = OUTPUT 2
RED LAMP = OUTPUT 3
YELLOW LAMP = OUTPUT 4

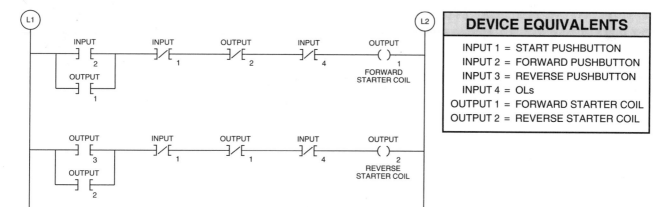

Programmable Controllers

Name _____ Date _____ **WORKSHEET 14-5**

Forward and Reversing PLC Diagram to Line Diagram Conversion

Use the Device Equivalents table to complete the standard line diagram.

1. Draw the standard line diagram of the basic PLC programming diagram.

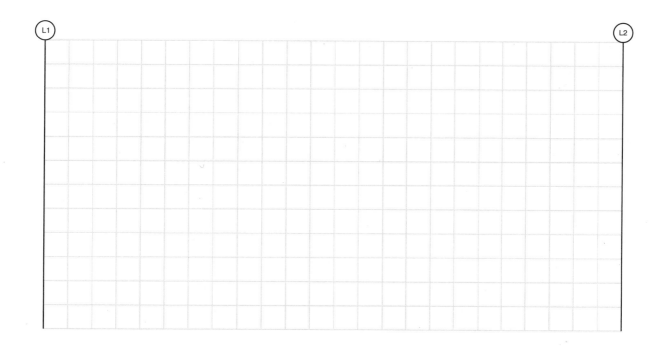

DEVICE EQUIVALENTS
INPUT 1 = START PUSHBUTTON
INPUT 2 = FORWARD PUSHBUTTON
INPUT 3 = REVERSE PUSHBUTTON
INPUT 4 = OLs
OUTPUT 1 = FORWARD STARTER COIL
OUTPUT 2 = REVERSE STARTER COIL

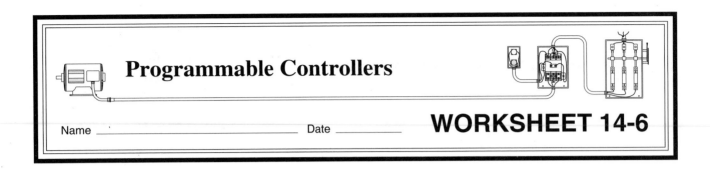

Programmable Controllers

WORKSHEET 14-6

Name _____ Date _____

Two-Speed PLC Diagram to Line Diagram Conversion

Use the Device Equivalents table to complete the standard line diagram.

1. Draw the standard line diagram of the basic PLC programming diagram.

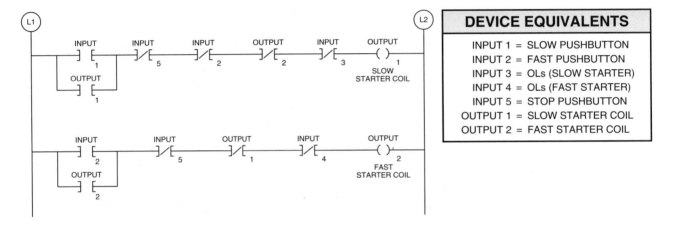

DEVICE EQUIVALENTS

INPUT 1 = SLOW PUSHBUTTON
INPUT 2 = FAST PUSHBUTTON
INPUT 3 = OLs (SLOW STARTER)
INPUT 4 = OLs (FAST STARTER)
INPUT 5 = STOP PUSHBUTTON
OUTPUT 1 = SLOW STARTER COIL
OUTPUT 2 = FAST STARTER COIL

Reduced-Voltage Starting

Name _____ Date _____

TECH-CHEK 15

Electrical Motor Controls

_____ 1. Reduced-voltage starting is used as a means of _____.

A. reducing starting current C. full-voltage starting
B. speed control D. starting difficult loads

_____ 2. The _____ is the part of a DC motor that is used to reverse the direction of current flow in the armature coils.

A. brush set C. commutator
B. power supply D. armature pole

_____ 3. The _____ is the part of a DC motor that is used to provide the contact to the external power supply.

A. brush set C. commutator
B. armature D. field pole

_____ 4. Sparking at the brushes of large DC motors is reduced by using _____.

A. commutators C. interpoles
B. additional main poles D. an insulator

_____ 5. A DC motor, unlike an AC motor, may need reduced-voltage starting to _____.

A. reduce starting current C. protect the electrical environment
B. reduce starting torque D. protect the motor

_____ 6. The speed of an AC squirrel cage motor can be changed by _____.

A. changing the number of poles C. changing the supply current
B. changing the supply voltage D. adding interpoles connected to the armature

_____ 7. Current and _____ of a motor are reduced when voltage is reduced to the motor.

_____ 8. In an AC squirrel cage motor, the fixed frame is the _____ and the rotating part is the rotor.

_____ 9. _____ current is the current required by a motor to produce full-load torque at the motor's rated speed.

_____ 10. _____ is the steady-state current taken from the power lines when a motor is started.

_____ 11. The five methods used in reduced-voltage starting are resistive, part-winding, wye-to-delta, solid-state, and _____.

_____ 12. Two reduced-voltage starting methods that are not adjustable to more than two steps without additional circuitry are the wye-to-delta and _____ starting methods.

_____ **13.** The reduced-voltage starting method which is adjustable through its entire range is the _____ starting method.

_____ **14.** _____ current is the maximum current permitted by the utility company in any one step of an increment start.

_____ **15.** A(n) _____ is the stationary windings or magnets of a DC motor.

_____ **16.** A(n) _____ is a flexible braided copper conductor used to connect the brushes to the external circuit.

_____ **17.** _____ is the effect that the magnetic field of the armature coil has on the magnetic field of the main pole windings.

_____ **18.** _____ is the percentage reduction in speed below synchronous speed.

_____ **19.** _____ starting uses a resistor connected in each motor line (in one line in a 1ϕ starter) to produce a voltage drop.

_____ **20.** _____ starting is a method of starting a motor by first applying power to part of the motor's coil windings for starting and then applying power to the remaining coil windings for normal running.

_____ **21.** _____ starting uses SCRs to control motor voltage, current, and torque during acceleration.

_____ **22.** _____ starting provides the highest possible starting torque per ampere of line current.

_____ **23.** _____ are auxiliary poles that are placed between the main field poles of a motor.

_____ **24.** As a DC motor accelerates, a _____ is generated which reduces the current in the motor.

_____ **25.** _____ motors have become the standard for AC, all-purpose, constant-speed motor applications.

Reduced-Voltage Starting

Name _____ Date _____

WORKSHEET 15-1

Primary Resistor Starting

Complete the wiring diagram according to the line diagram. Do not make any wire splices or additional terminal connections on the wiring diagram. All connections must run from terminal screw to terminal screw.

1. Design a circuit so when the start pushbutton is pressed, the contactor (M) and the time-delay relay (TR) are energized and the motor is connected to the incoming power lines through the resistor bank. After the time-delay relay has timed out, the timed closed contacts close and contactor A1 is energized, shorting across each of the three resistors in the resistor bank. This automatically switches the motor to full power.

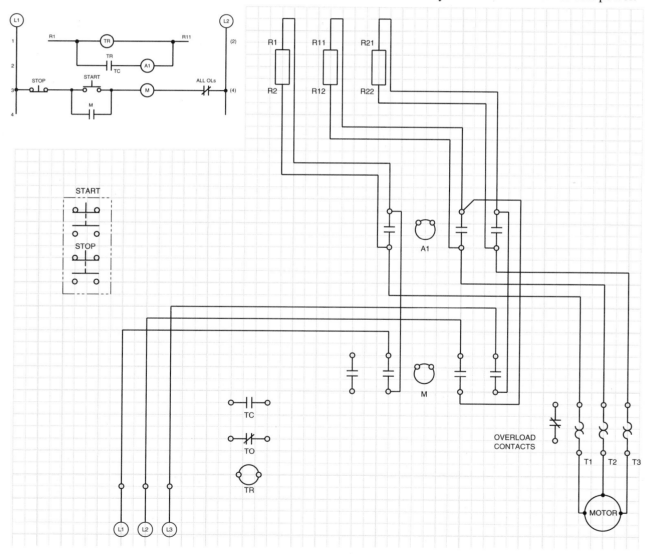

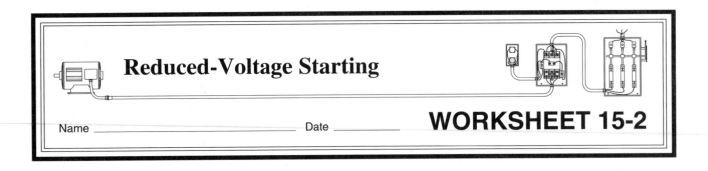

Reduced-Voltage Starting

Name _____ Date _____

WORKSHEET 15-2

Part-Winding Starting

Complete the circuits based on the given information. Do not make any wire splices or additional terminal connections on the wiring diagram. All connections must run from terminal screw to terminal screw.

1. Design a circuit so when the start pushbutton is pressed, contactor M1 is energized first and power is applied to motor terminals T1, T2, and T3. After the time-delay, the NO auxiliary interlock on M1 times out, and the timed closed (TC) contacts close, and the M2 contactor is energized, connecting power to the second winding (motor terminals T7, T8, and T9). The motor is stopped by pressing the stop pushbutton, which drops out both contactors. If motor terminals T4, T5, and T6 are not internally connected, they should be wired together at the terminal box as indicated by the dotted power lines at the motor.

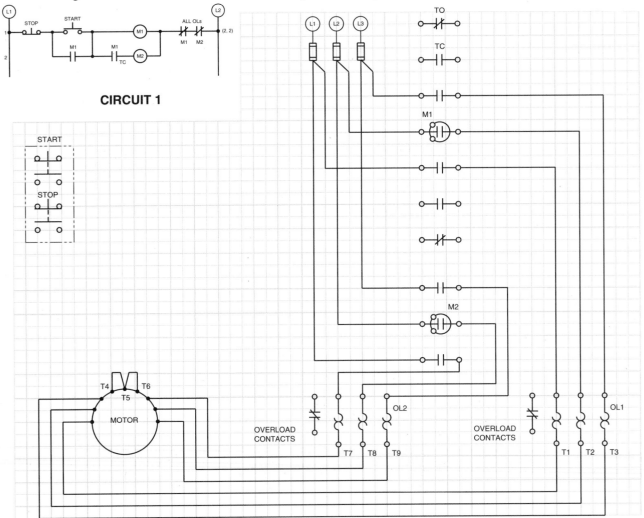

CIRCUIT 1

2. Redraw the line diagram of Circuit 1, adding a separate timer to the circuit that is used when a time-delay auxiliary contact is not used.

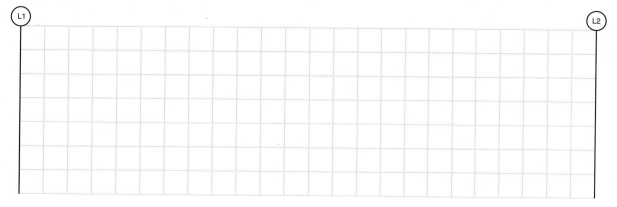

CIRCUIT 2

3. Redraw the line diagram of Circuit 1, adding a second timer that does not allow the motor to be restarted until it has stopped for 10 min. Replace the pushbutton station with a temperature controller because this type of operation is normally automatic.

4. Redraw Circuit 2, adding a cooling fan motor to run for 10 min after the reduced-voltage starting motor has turned OFF.

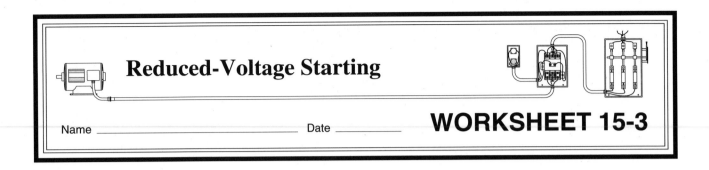

Wye/Delta Connections

Complete the wiring diagram according to the line diagram. Do not make any wire splices or additional terminal connections on the wiring diagram. All connections must run from terminal screw to terminal screw.

1. Design a circuit so when the start pushbutton is pressed, contactors S and M1 are energized. Contactor S connects motor terminals T4, T5, and T6. Contactor M1 connects the incoming power lines to motor terminals T1, T2, and T3, causing the motor to start in a wye-connected configuration. After the time-delay NC interlock on M1 times out, the timed open (TO) contacts open, dropping out contactor S and picking up contactor M2. The M2 contactor applies power to terminals T4, T5, and T6, bringing the motor up full speed in a delta-connected configuration. The motor is stopped by pressing the stop pushbutton, which drops out all three contactors.

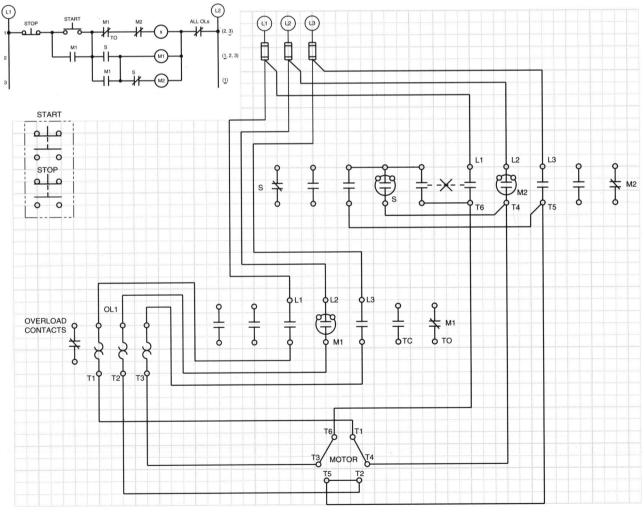

Reduced-Voltage Starting

Name _____ Date _____

WORKSHEET 15-4

Autotransformer Starting

Complete the wiring diagram according to the line diagram. Do not make any wire splices or additional terminal connections on the wiring diagram. All connections must run from terminal screw to terminal screw.

1. Design a circuit so when the start pushbutton is pressed, the time-delay relay (TR) and contactors S and M are energized, applying power through the windings of the autotransformer to the motor. When the time delay relay times out, the timed open (TO) contacts open and the timed closed (TC) contacts close, the S contactor drops out, and the R contactor is energized, switching the motor to full-line voltage. The autotransformer incorporates a thermal protector switch imbedded in the winding of each of the two transformer coils. These devices (TPA and TPB) sense the heat rise in the coils and open their NC contacts if the temperature limits are reached. This allows full current to the lockout relay (LR) and opens its NC contacts. LR is normally shorted out by the thermal protector switches. The lockout relay has to be hand-reset to restore power to the line.

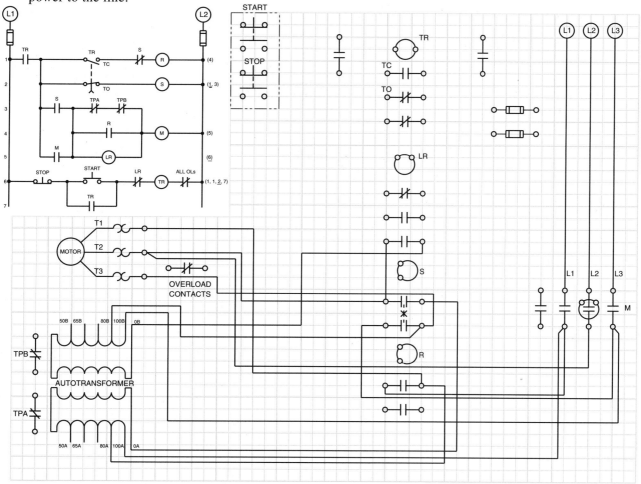

Reduced-Voltage Starting

Name _____ Date _____

WORKSHEET 15-5

Motor Load Monitoring

Use Data Sheet L to complete the line diagram.

1. Draw the line diagram showing how a load guard power factor monitor may be added to a standard start/stop pushbutton circuit. The start/stop station controls the grinder and the load guard automatically stops the motor when a jam or overload occurs.

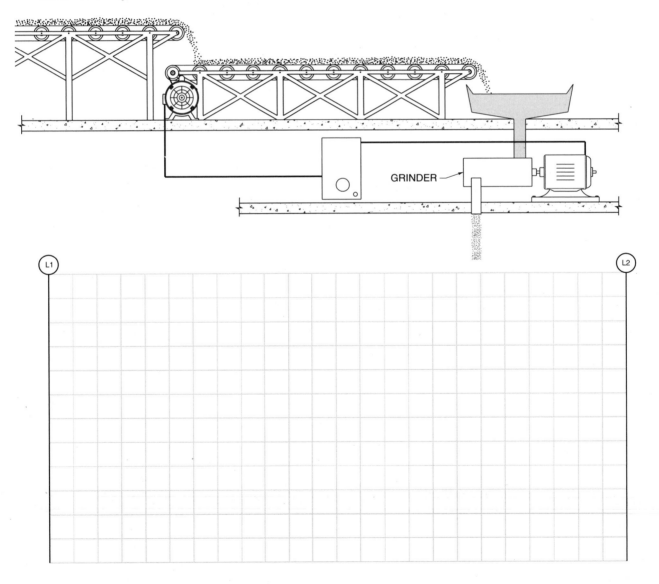

GRINDER

L1   L2

Accelerating and Decelerating Methods

Name _____ Date _____

TECH-CHEK 16

Electrical Motor Controls

_____ **1.** Braking torque developed with friction brakes is directly proportional to the _____.

A. speed of the motor   C. applied voltage
B. surface area and spring pressure   D. type of solenoid used

_____ **2.** The advantage of using friction brakes is _____.

A. their ability to be connected to any   C. their ability to control loads that are
voltage   often stopped
B. simplified maintenance and   D. low maintenance
lower initial cost

_____ **3.** In electric braking, the amount of braking force is varied by changing the _____.

A. spring pressure   C. surface area of the brake
B. time the motor is connected   D. applied DC voltage
in reverse

_____ **4.** Dynamic braking is normally applied to DC motors because _____.

A. DC motors reverse faster   C. AC motors do not have stationary
than AC motors   fields
B. there must be access to the   D. AC power cannot develop a retarding
rotor   torque

_____ **5.** Friction brakes are sometimes used with dynamic braking because _____.

A. friction brakes are less expensive   C. dynamic braking uses too much power
B. dynamic braking cannot hold   D. AC brakes cannot be used with DC
a stopped load

_____ **6.** A _____ is a load that requires a constant torque/variable horsepower motor.

A. paper roll machine   C. fan
B. clock   D. conveyor

_____ **7.** A _____ is a load that requires a constant horsepower/variable torque motor.

A. paper roll machine   C. fan
B. clock   D. conveyor

_____ **8.** A _____ is a load that requires a variable torque/variable horsepower motor.

A. paper roll machine   C. fan
B. clock   D. conveyor

_____ **9.** A(n) _____-delay timer is used for a timing relay to plug a motor to a stop.

_____ **10.** A motor used for plugging should have a service factor of _____ or more.

_____ **11.** Electric braking is a method of braking in which a(n) _____ current is applied to the
stationary windings of a motor after the voltage is removed.

_____ **12.** In dynamic braking, the smaller the resistance of the resistor used, the _____ the rate of energy dissipation.

_____ **13.** _____ is the force that produces or tends to produce rotation in a motor.

_____ **14.** _____ is the torque a motor produces when the rotor is stationary and full power is applied to the motor.

_____ **15.** Electric power is rated in horsepower or _____.

_____ **16.** The frequency or _____ must be changed to change the speed of an AC induction motor.

_____ **17.** An AC induction motor with four poles runs at a synchronous speed of _____ rpm.

_____ **18.** _____ circuit logic is a circuit that starts a motor in low speed and automatically brings the motor to high speed after the high pushbutton is pressed.

_____ **19.** _____ circuit logic is a circuit that requires a motor to start at low speed before the motor can be put into high speed.

_____ **20.** A(n) _____ changes standard 60 Hz AC power into almost any desired frequency.

_____ **21.** A(n) _____ changes DC voltage into AC variable frequency.

_____ **22.** A driven machine (load) rotates at a(n) _____ speed than the motor if a large pulley is placed on the driven machine (load) and a small pulley is placed on the motor.

_____ **23.** _____ are used in very low-speed applications.

_____ **24.** The work required to move a 150 lb load 10′ is _____ lb-ft.

_____ **25.** A total of _____ lb-ft of torque is produced by a 75 lb force pushing on a 2′ lever arm.

_____ **26.** The full-load torque of a ½ HP motor operating at 1725 rpm is _____ lb-ft.

_____ **27.** An increase in a motor's speed causes a(n) _____ in the motor's horsepower output if the torque requirements on the motor remain constant.

_____ **28.** An increase in a motor's torque requirement causes a(n) _____ in the motor's horsepower output if the speed remains constant.

_____ **29.** A total of _____ lb-ft of torque is produced by a 100 lb force pushing on a 4′ lever arm.

_____ **30.** The full-load torque of a 5 HP motor operating at 1725 rpm is _____ lb-ft.

_____ **31.** The full-load torque of a 5 HP motor operating at 1140 rpm is _____ lb-ft.

_____ **32.** A 460 V, 85% efficient motor pulling 6 A produces _____ HP.

_____ **33.** A 230 V, 85% efficient motor pulling 12 A produces _____ HP.

_____ **34.** The synchronous speed of a 2-pole motor operating at 50 Hz is _____ rpm.

_____ **35.** A(n) _____″ driven machine pulley diameter is required if a motor running at 1200 rpm has a 3″ pulley and the driven machine is running at 600 rpm.

_____ **36.** A(n) _____″ driven machine pulley diameter is required if a motor running at 1200 rpm has a 3″ pulley and the driven machine is running at 2400 rpm.

Accelerating and Decelerating Methods

Name _____   Date _____

WORKSHEET 16-1

Motor Plugging

Complete the wiring diagram based on the line diagram. Do not make any wire splices or additional terminal connections on the wiring diagram. All connections must run from terminal screw to terminal screw.

1. The motor is started by pushing the start pushbutton. As it accelerates, centrifugal force closes the NO contacts of the plugging switch. This closes relay CR1, which is held closed by its NO contacts. The forward contactor (F) opens, closing the reverse contactor (R) through its NC auxiliary contacts and through the contacts of relay CR1 when the stop pushbutton is pressed. The reverse contactor applies reverse power to the motor until its speed is reduced to allow the contacts of the plugging switch to open. The opening of the contacts causes CR1 and R to open, disconnecting the motor from the line. Relay CR1 is a safety interlock that prevents the starter from energizing when the motor shaft is turned by hand.

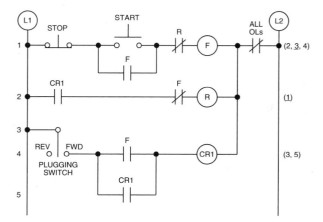

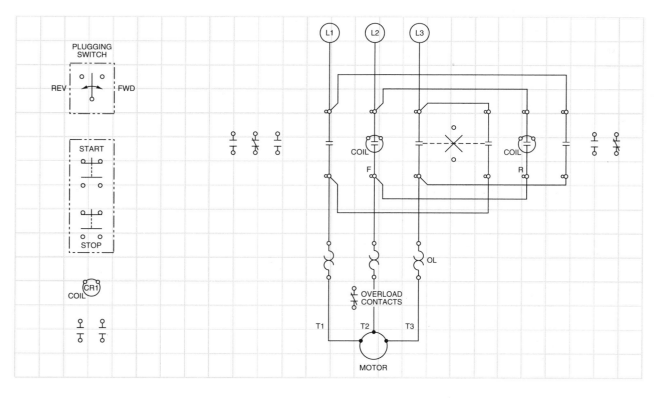

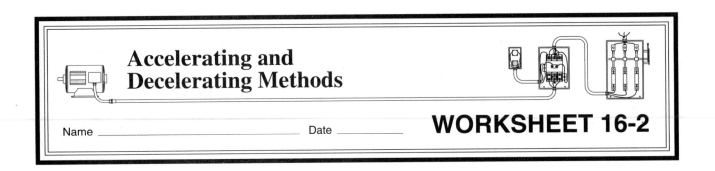

Accelerating and Decelerating Methods

Name _____ Date _____

WORKSHEET 16-2

Two-Speed Starting

Complete the wiring diagram based on the line diagram. Do not make any wire splices or additional terminal connections on the wiring diagram. All connections must run from terminal screw to terminal screw.

1. The circuit provides for compelling, two-speed starting with two thermal overload relays, one for each speed. The controller is internally wired to compel the operator to start the motor at slow speed. It cannot be switched to fast speed until after the motor is running. The slow-speed contactor and the control relay (FR) are energized when the slow pushbutton is pressed. The motor does not start if the fast pushbutton is pressed because the NC contacts of the control relay prevent the high-speed contactor from energizing. Once the motor is running, pressing the fast pushbutton automatically opens the slow-speed contactor and closes the high-speed contactor through the NC contacts of the slow-speed contactor and the NO contacts of the control relay. The starter cannot be switched from fast to slow without first pressing the stop pushbutton.

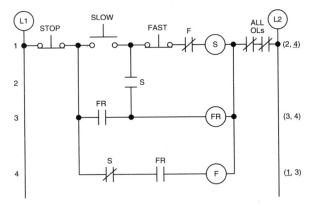

MOTOR CONNECTIONS			
SPEED	LINES		
	L1	L2	L3
SLOW	T1	T2	T3
FAST	T11	T12	T13

Accelerating and Decelerating Methods

Name _____ Date _____

WORKSHEET 16-3

Multispeed Starting

Complete the wiring diagram based on the line diagram. Do not make any wire splices or additional terminal connections on the wiring diagram. All connections must run from terminal screw to terminal screw.

1. The circuit provides for multispeed starting using a constant-horsepower motor. The motor control is a three-element fast/slow/stop pushbutton station connected for starting in either the fast or slow speed. The stop pushbutton must be pressed first to change speed from fast to slow.

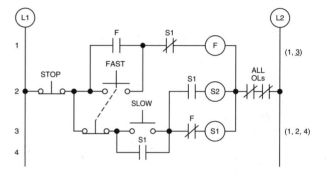

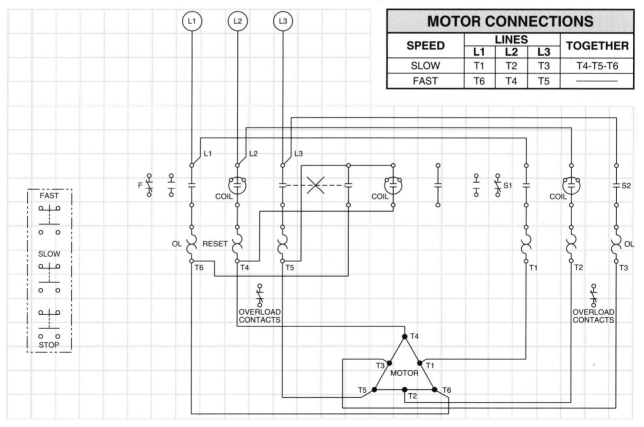

MOTOR CONNECTIONS				
SPEED	**LINES**			**TOGETHER**
	L1	**L2**	**L3**	
SLOW	T1	T2	T3	T4-T5-T6
FAST	T6	T4	T5	——

Accelerating and Decelerating Methods

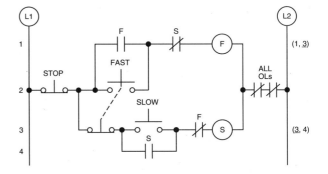

Name _____ Date _____ **WORKSHEET 16-4**

Selective Multispeed Starting

Complete the wiring diagram based on the line diagram. Do not make any wire splices or additional terminal connections on the wiring diagram. All connections must run from terminal screw to terminal screw.

1. The circuit provides for selective, multispeed starting with two thermal overload relays. The control is a three-element fast/slow/stop pushbutton station. The motor can be started at fast or slow speed but cannot be switched from fast to slow without first pressing the stop pushbutton.

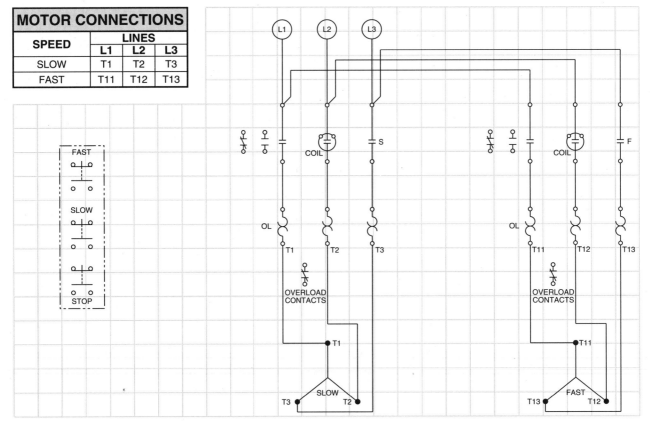

MOTOR CONNECTIONS			
SPEED	LINES		
	L1	L2	L3
SLOW	T1	T2	T3
FAST	T11	T12	T13

Accelerating and Decelerating Methods

Name _____ Date _____

WORKSHEET 16-5

Unrestrictive Multispeed Starting

Complete the wiring diagram based on the line diagram. Do not make any wire splices or additional terminal connections on the wiring diagram. All connections must run from terminal screw to terminal screw.

1. The motor control is a standard three-element fast/slow/stop pushbutton station connected for starting at either fast or slow speed. It cannot be switched from fast to slow without first pressing the stop pushbutton.

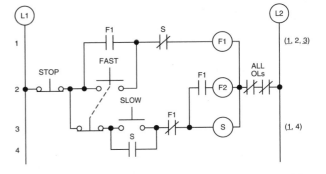

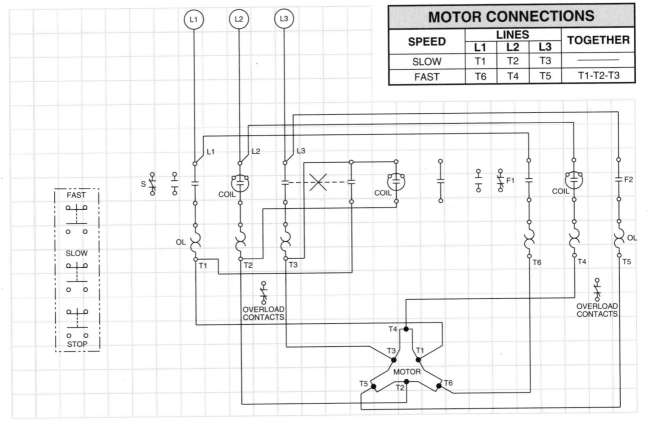

MOTOR CONNECTIONS				
SPEED	LINES			TOGETHER
	L1	L2	L3	
SLOW	T1	T2	T3	—
FAST	T6	T4	T5	T1-T2-T3

Accelerating and Decelerating Methods

Name _____ Date _____

WORKSHEET 16-6

Compelling Circuit Logic

Design a line diagram for compelling circuit logic using three motor starters (low, medium, and high). Use standard lettering, numbering, and coding information.

1. The motor must be started at low speed before changing to medium speed and must be running at medium speed before changing to high speed. Use two control relays to complete the circuit. Provide overload protection at all speeds.

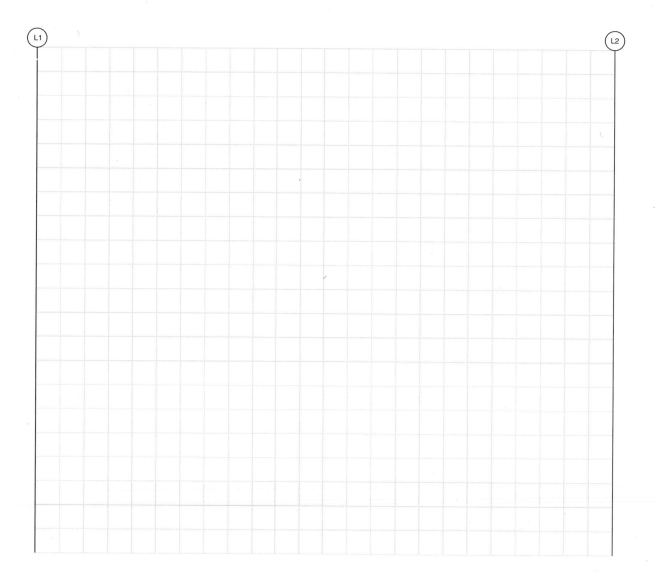

Accelerating and Decelerating Methods

Name _____ Date _____

WORKSHEET 16-7

Dynamic Braking

Design a circuit in which a friction brake can be applied to a DC shunt motor wired for dynamic braking. Use standard lettering, numbering, and coding information.

1. The friction brake is to be applied 3 sec after the motor is turned OFF, allowing the dynamic braking action to first slow the motor. The friction brake is to remain ON, holding the load until the motor is started again.

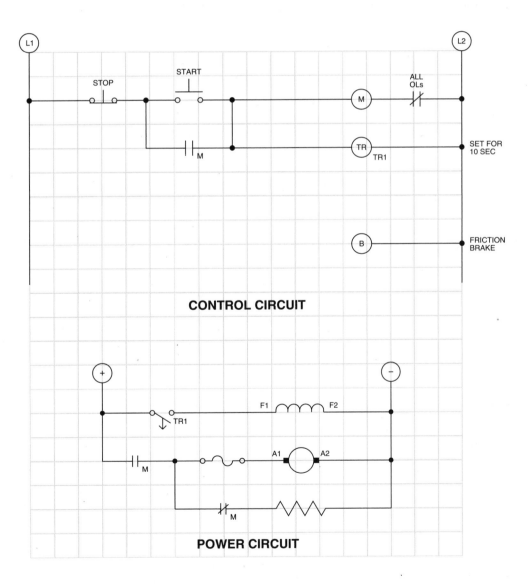

Accelerating and Decelerating Methods

Name _____ Date _____

WORKSHEET 16-8

Motor Temperature Control

Design a circuit in which a temperature switch turns ON and OFF a heating element and circulating fan based on a given temperature. Use standard lettering, numbering, and coding information.

1. Include a plugging switch which automatically turns OFF the heating element contactor if the circulating fan is not operating. Provide overload protection for the motor.

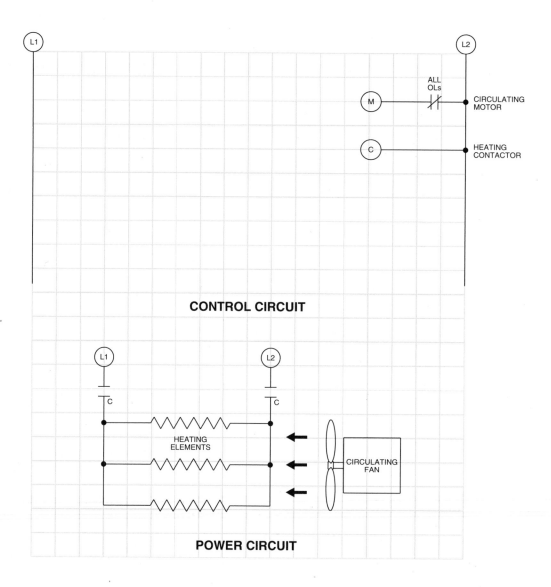

CONTROL CIRCUIT

POWER CIRCUIT

Preventive Maintenance and Troubleshooting

Name _____ Date _____

TECH-CHEK 17

Electrical Motor Controls

_____ 1. Phase unbalance in a 3φ power system normally occurs because _____ the system.

    A. 3φ loads are removed from   C. 1φ loads are added to
    B. 3φ loads are added to   D. 1φ loads are removed from

_____ 2. A _____ motor continues to run with a phase loss.

    A. 1φ   C. DC
    B. 3φ   D. capacitor-run

_____ 3. The NEC® requires protection against _____ on all equipment transporting people.

    A. phase loss   C. phase unbalance
    B. phase reversal   D. single-phasing

_____ 4. _____ is the major cause of large voltage surges on power lines.

    A. Frequency variation   C. A 3φ motor being turned OFF
    B. A lightning strike   D. A 1φ motor being turned OFF

_____ 5. Motor speed _____ when the frequency of the power lines operating an AC motor is decreased.

    A. remains the same   C. decreases
    B. increases   D. A, B, and C

_____ 6. The power limit of a transformer is listed on the nameplate of the transformer under the _____ rating.

    A. voltage   C. secondary
    B. current   D. kVA

_____ 7. The major cause of most motor failures is _____.

    A. capacitor failure   C. bearing failure
    B. mechanical breakage   D. the deterioration of the insulation

_____ 8. The needle on an ohmmeter _____ to indicate a good capacitor.

    A. swings to zero and does not move   C. swings to zero and slowly moves to infinity
    B. remains on infinity   D. swings to infinity and slowly moves to zero

_____ 9. A(n) _____ test is used to check motor insulation over the life of the motor.

    A. insulation spot   C. insulation step voltage
    B. dielectric absorption   D. phase-loss

_____ 10. A(n) _____ test is used to check for moisture and contamination on insulation.

    A. insulation spot   C. insulation step voltage
    B. dielectric absorption   D. phase-loss

_____ **11.** The series field in a DC motor normally has a reading _____ the armature reading.

    A. greater than   C. the same as
    B. less than   D. cannot be determined

_____ **12.** The shunt field normally has a reading _____ the armature reading.

    A. greater than   C. the same as
    B. less than   D. cannot be determined

Voltage Identification

_____ **1.** Half-wave DC

_____ **2.** Full-wave DC

_____ **3.** 1ϕ

_____ **4.** 3ϕ

_____ **5.** Pure DC

_____ **6.** Semifiltered DC

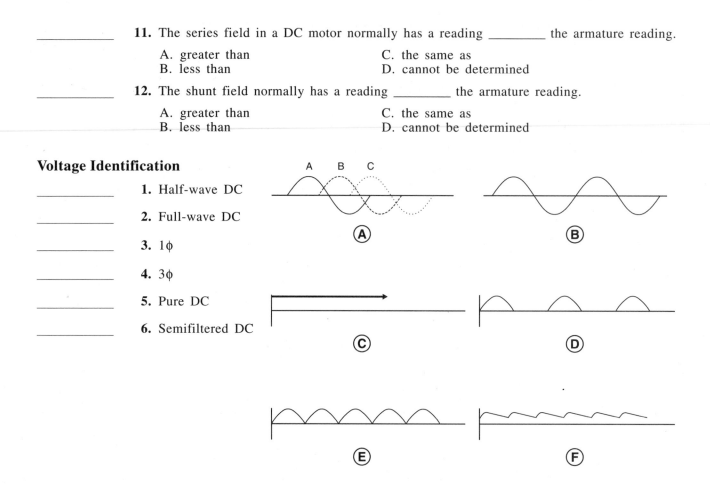

Motor Insulation Tests

_____ **1.** Curve _____ indicates good motor insulation.

_____ **2.** Curve _____ indicates good motor insulation.

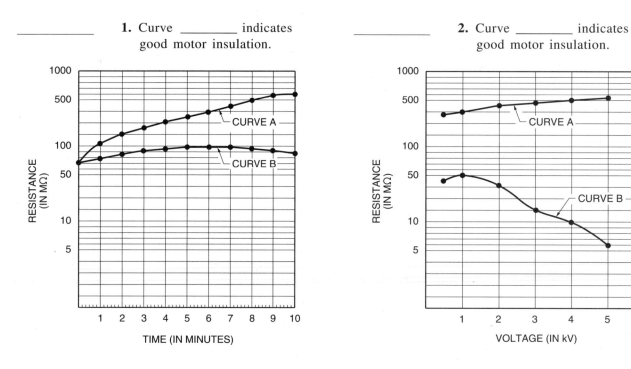

Preventive Maintenance and Troubleshooting

Name _____   **Date** _____

WORKSHEET 17-1

Motor Protection

Use Data Sheet M to complete the wiring diagram and line diagram.

1. Design a circuit using the start/stop pushbutton station and a Model M relay to protect against phase loss or phase angle error (adjustable from 5° to 15°). Phase angle error takes place in an unbalanced system. The relay is rated for the same voltage as the power lines. Connect the relay so it adds protection along with the motor starter overloads.

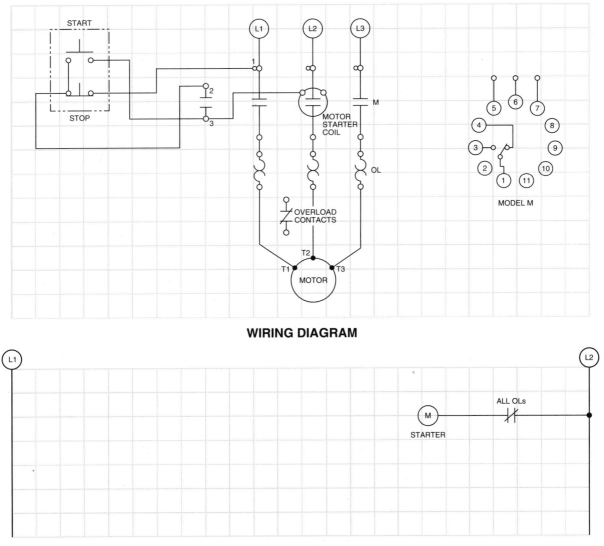

WIRING DIAGRAM

LINE DIAGRAM

Preventive Maintenance and Troubleshooting

Name _____ Date _____ **WORKSHEET 17-2**

Phase Loss Protection

Use Data Sheet N to complete the wiring diagram and line diagram.

1. Design a circuit using the start/stop pushbutton station and a Model N relay to de-energize the motor starter if one or more phases is lost. The relay also de-energizes the motor starter if the phase sequence is not correct. L1 (R) must be connected to pin 5. L2 (S) must be connected to pin 6. L3 (T) must be connected to pin 7. Any other sequence de-energizes the relay. Connect the relay so it adds protection along with the overload contacts.

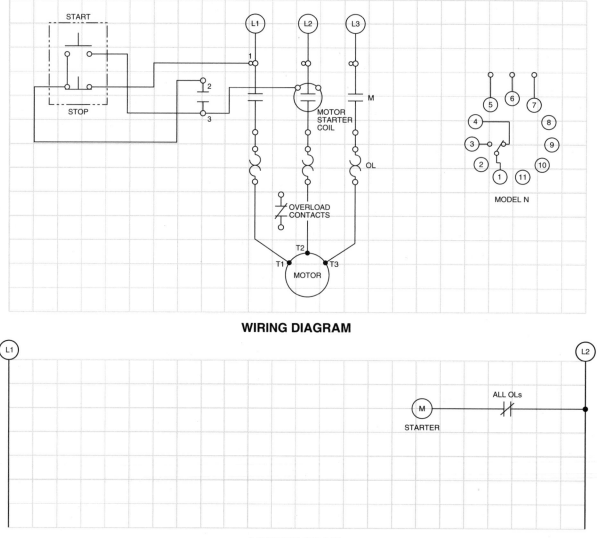

WIRING DIAGRAM

LINE DIAGRAM

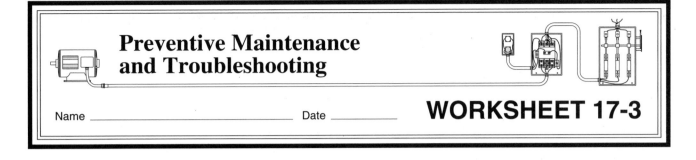

Preventive Maintenance and Troubleshooting

Name _____ Date _____

WORKSHEET 17-3

Overload Protection

Use Data Sheet O to complete the wiring diagrams and line diagrams.

1. Design a circuit using the start/stop pushbutton station and a Model O relay to protect against an overload in the power circuit. The relay can be used on 1φ or 3φ circuits. It uses a current transformer that detects the amount of current in any wire that passes through the current transformer. It can be used with or in place of the standard overload heater found on magnetic motor starters. The exact setting of the relay, like the selection of overload heaters, must meet NEC® requirements. The maximum setpoint of the relay is normally 1.15 or 1.25 times the FLC which is found on the nameplate of the motor. For exact requirements, see Article 430 of the NEC®.

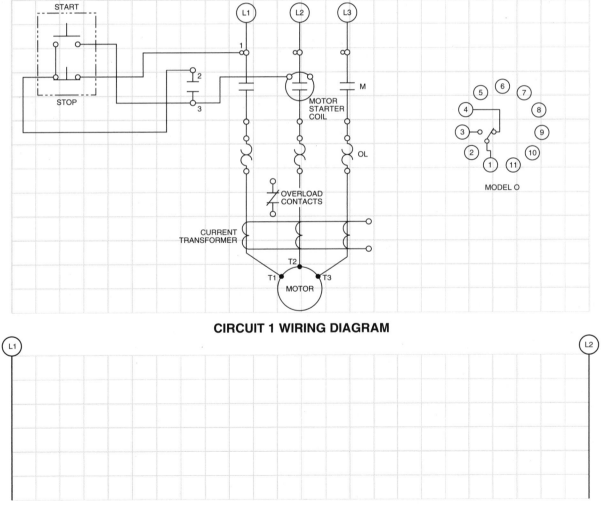

CIRCUIT 1 WIRING DIAGRAM

CIRCUIT 1 LINE DIAGRAM

2. Redraw the line diagram of Circuit 1, adding a timer to allow the locked rotor current to exist for 5 sec before the relay monitors the full-load current. This is required because the relay reacts instantly to an overload current and turns OFF the motor. This does not allow the locked rotor current required to start the motor.

CIRCUIT 2 LINE DIAGRAM

3. Redraw the line diagram of Circuit 2, adding a second timer to allow a full-load overload current for 10 sec before the motor is automatically turned OFF. Thus, timer 1 allows the locked rotor current (starting current) to exist for 5 sec and timer 2 allows the full-load current (running current) to exist for 10 sec before the motor is automatically turned OFF.

Preventive Maintenance and Troubleshooting

Name _____ Date _____

WORKSHEET 17-4

Manual/Automatic Circuit Troubleshooting

Troubleshoot the circuit based on the given information.

1. The solenoid energizes when the pushbutton or pressure switch is closed. The pilot light does not turn ON when the foot switch and temperature switch close. Add a fused jumper wire to the circuit to eliminate trouble with the control switches. Assume that the light does not light when the jumper wire is in place. Circle the part of the circuit that contains the fault.

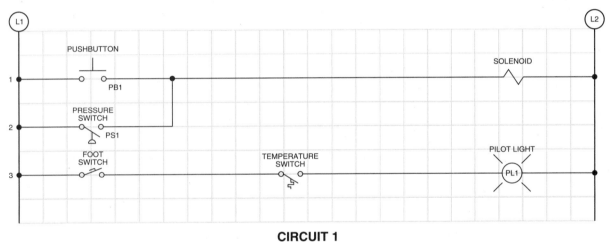

CIRCUIT 1

2. Redraw Circuit 1, adding a pushbutton that can be used to test PL1 any time it is pressed.

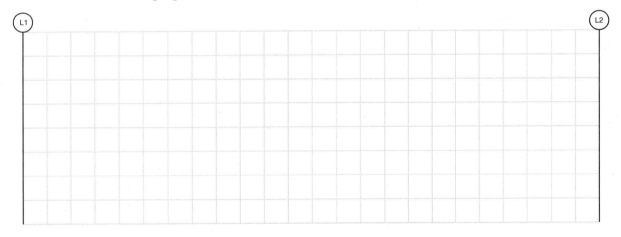

Preventive Maintenance and Troubleshooting

Name _____ Date _____

WORKSHEET 17-5

Carton Fill Circuit Troubleshooting

Troubleshoot the circuit based on the given information.

1. The empty cartons are stopping and staying in place for the given amount of time when they hit the limit switch. The cartons are not being filled. Add a voltmeter to the circuit to test the solenoid valve. Assume that the voltmeter indicates a proper voltage reading at the correct time. Circle the part of the circuit that contains the fault.

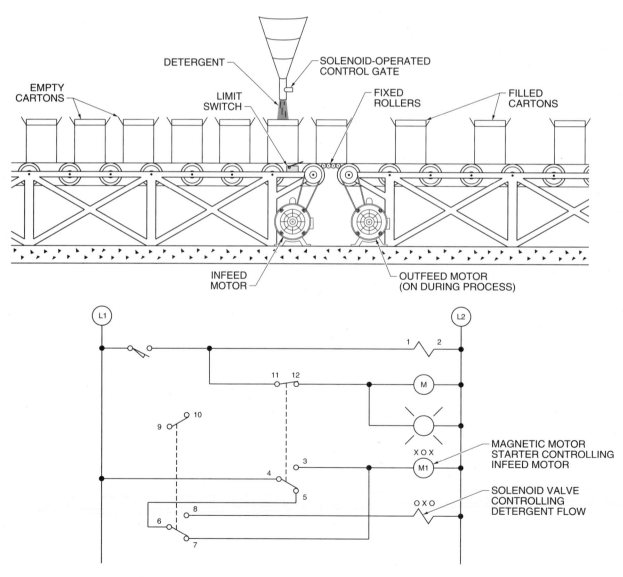

Preventive Maintenance and Troubleshooting

Name _____ Date _____

WORKSHEET 17-6

Multiple Starter Circuit Troubleshooting

Troubleshoot the circuit based on the given information.

1. Warning light PL2 is ON. A check of starting coil M2 indicates that the starter is energized and the motor is running. Circle the part of the circuit that contains the fault.

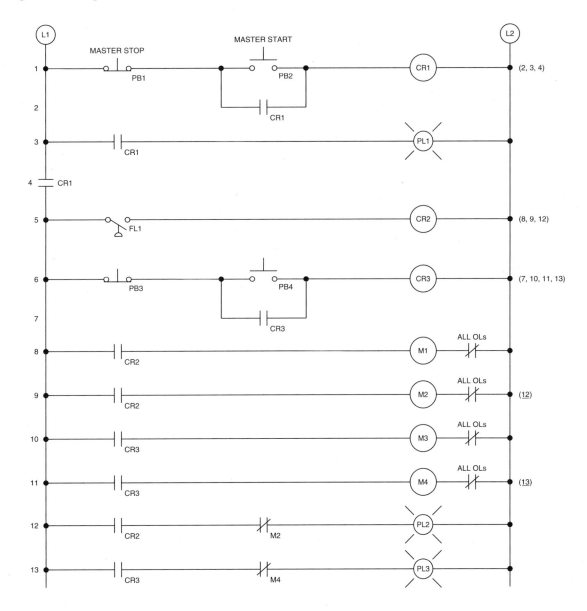

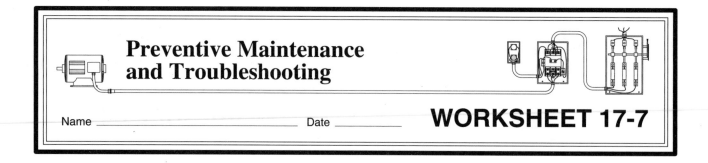

Preventive Maintenance and Troubleshooting

Name _____ Date _____

WORKSHEET 17-7

Primary Resistor Starting Circuit Troubleshooting

Troubleshoot the circuit based on the given information.

1. The motor is running hot and does not seem to have much power. A test with a voltmeter indicates that there is only about ½ the required voltage at terminals T1, T2, and T3 of the motor. Add a fused jumper(s) to eliminate trouble with the control circuit. Connect the voltmeter to test the power circuit for the source of trouble.

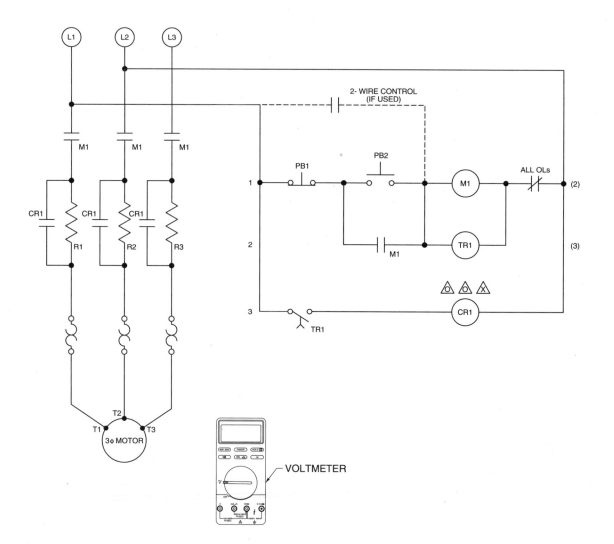

**Preventive Maintenance
and Troubleshooting**

Name _____   Date _____

WORKSHEET 17-8

Motor Braking Circuit Troubleshooting

Troubleshoot the circuit based on the given information.

1. The motor is not braking to a stop. A test of the brake contactor in the control circuit indicates that the contactor is energizing at the correct time. Connect voltmeter 1 to test for the correct AC output from the transformer. Connect voltmeter 2 to test for the correct DC output when the brake contactor is energized.

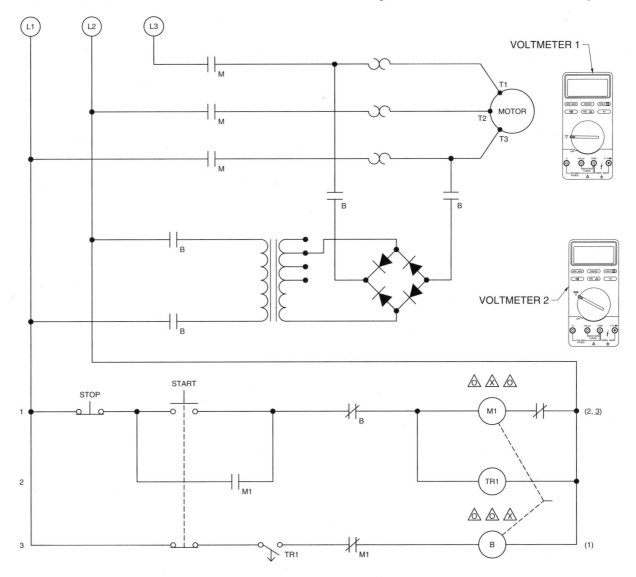

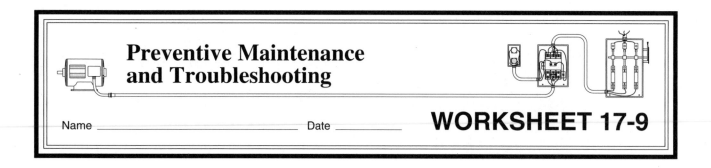

Preventive Maintenance and Troubleshooting

Name _____ Date _____

WORKSHEET 17-9

Selector Switch Circuit Troubleshooting

Troubleshoot the circuit based on the given information.

1. Magnetic starter coil M1 starts and remains engaged after the start pushbutton is pressed regardless of the position of the selector switch. Circle the part of the circuit that contains the fault.

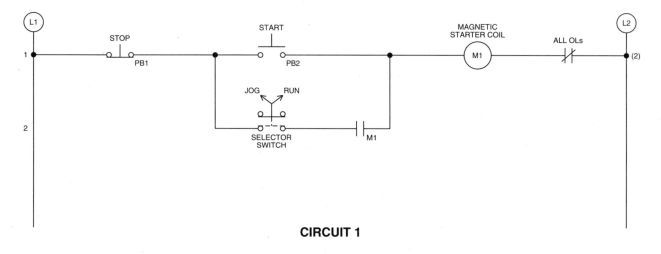

CIRCUIT 1

2. Redraw Circuit 1, adding a red lamp that is ON any time the selector switch is in the run position and a yellow lamp that is ON any time the selector switch is in the jog position. Only one lamp is ON at a time.

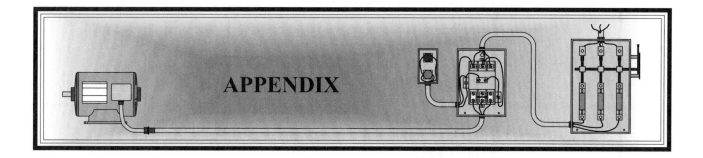

APPENDIX

Data Sheet A

Heating Element Connections

A common dual-element heating coil may be used in most electric ranges. The schematic diagram shows, with graphic symbols, the electrical connections and functions of a specific circuit arrangement. A schematic diagram does not show the physical relationship of the components in a circuit. See Figure A1.

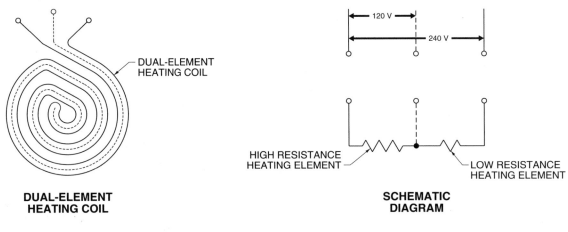

DUAL-ELEMENT HEATING COIL

SCHEMATIC DIAGRAM

FIGURE A1

As many as 10 different temperature ranges are possible with two elements and a 120/240 V power supply, depending on how the elements and the power supply are connected. See Figure A2. Connections to the power supply may be made in series, parallel, or series/parallel combinations.

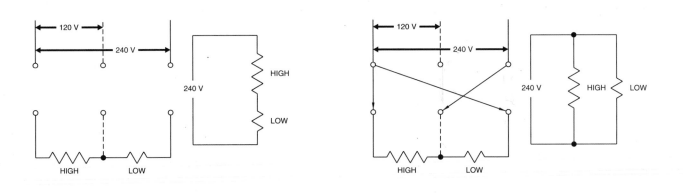

FIGURE A2

Wire Numbers

A control circuit may consist of a few individual wires or several thousand interconnected wires. Tracing one wire may be impossible without a system for keeping track of each wire in the circuit. The standard industrial wiring numbering system gives each wire or common group of wires a number. A common group is any wires that are connected directly without being broken by a device such as a pushbutton, contact, or starting coil.

Numbers are assigned in ascending numerical order (1, 2, 3, etc.), beginning with the upper left corner and moving to the right, line by line. A new number is assigned whenever a wire (or common group of wires) is broken by an electric device. See Figure B1.

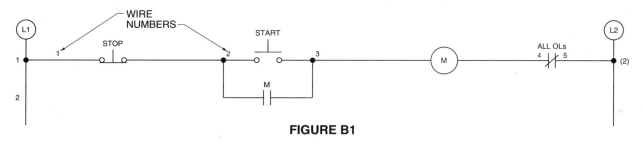

FIGURE B1

The first wire is assigned 1 and is referred to as wire 1 until it is broken by the stop pushbutton. After the stop pushbutton, wire 2 begins and continues until it is broken by the start pushbutton and the M contact. The wire going to the start pushbutton and the wire going to the M contact are referred to as wire 2 because wire 2 has not been broken by an electric device.

After the start pushbutton and the M contact, both wires are referred to as wire 3 until broken by the M starting coil. Wire 4 is assigned after the starting coil and continues until it is broken by the overload contacts. Wire 5 is the wire from the overload contacts to L2.

As a circuit's complexity increases, the numbers continue to ascend from the top left to the bottom right, line by line. See Figure B2. In this circuit, seven numbers are assigned. Some are assigned to individual wires and some to common groups of wires. A new number is assigned each time the interconnected wires are broken by an electric device.

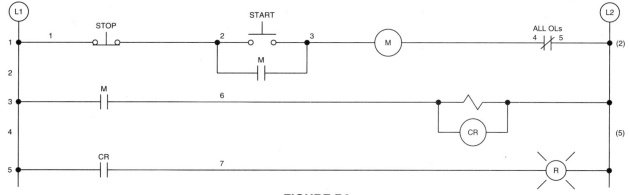

FIGURE B2

Wire Numbering System Simplification

In Figures B1 and B2, each wire was assigned a number regardless of its purpose or location in the circuit. Although this procedure is followed for some circuits, most circuits follow a simplified numbering system. In the simplified numbering system, any wire prewired by the manufacturer prior to shipping is not numbered. For example, the wire connecting the motor starting coil to the overload contacts is normally prewired and does not require an assigned number. See Figure B3.

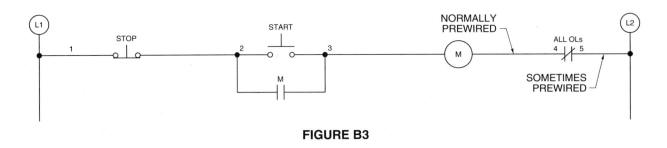

FIGURE B3

The numbering system may be further simplified by not numbering the wire directly connected to L2. This is because the wire from the overload contact to L2 is often prewired. The wire from L1 is always numbered.

The numbering system used for a given circuit applies to that circuit only. Numbering can differ from one circuit to another, even if the circuits are electrically identical. For example, three circuits that are electrically identical may be numbered differently. See Figure B4.

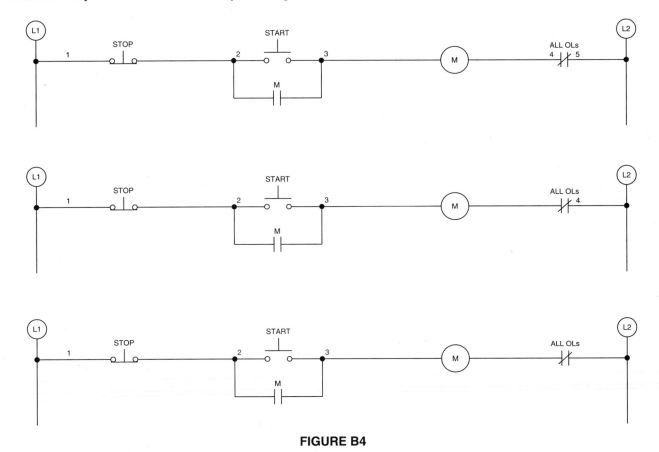

FIGURE B4

Wire Numbering System Advantages

The use of a wire numbering system simplifies initial circuit wiring and subsequent troubleshooting. The need for such a system becomes apparent when working on a wiring diagram in which wires enter and exit the conduit and are often grouped with many other wires.

Another advantage of a wire numbering system is that its use by manufacturers allows them to provide illustrations of connections for different wiring combinations. See Figure B5. The wiring diagram is provided by the motor starter manufacturer when the motor starter is purchased with an enclosure. The manufacturer includes the diagram on the inside cover of the enclosure. All major manufacturers use the standard wire numbering system described, with only slight variations.

In Figure B5, the manufacturer has prewired the motor coil to one side of the overload contacts. The other side of the overload contacts is connected directly to L2 and is marked X2. The separate control note explains why the contact is marked X2.

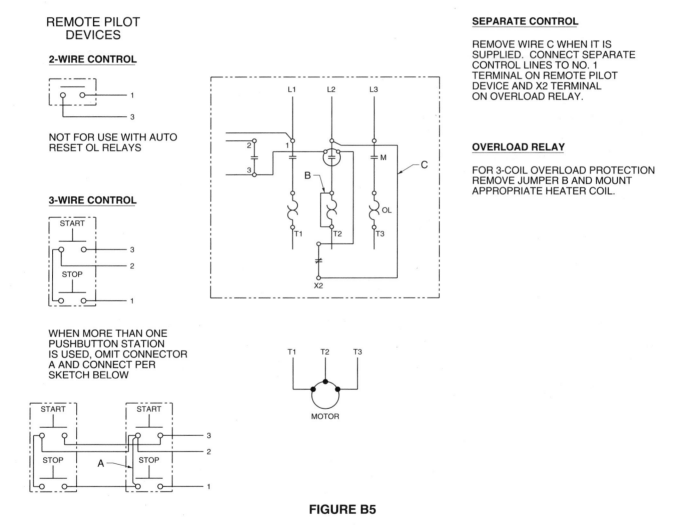

FIGURE B5

The prewiring is left alone if the starting coil is the same voltage level as the supply lines (L1 and L2). Wire C is removed from L2 if the starting coil is not the same voltage level as the supply lines, or if a separate control supply is used. For example, if a 120 V starting coil controls a 240 V motor, L1 and L2 must be 240 V to power the motor. In this case, wire C must be removed and connected to a 120 V supply. For this reason, X2 is marked on the overload indicating a connection to a lower supply voltage.

In Figure B5, the manufacturer has marked wire points 1, 2, and 3 on the diagram. These numbers are standard and are marked directly on all starters from major manufacturers. The starter may be wired for 2-wire, 3-wire, or multiple-control stations with each point marked in this manner. For example, if a 2-wire device, such as a pressure, float, or temperature switch, is used to control the motor, the control device is connected to points 1 and 3. If a 3-wire control device, such as a pushbutton station, is used to control the motor, it is connected to points 1, 2, and 3.

As circuits become more complicated, it is impossible for manufacturers to apply a standard numbering system to every circuit. Instead, the electrician must number the wires in each circuit at the time of wiring. An exception is the circuits that are used frequently, such as circuits that forward and reverse a motor with a standard start/stop pushbutton station. In this case, the manufacturer may provide the wire numbering.

Line Diagrams Used to Connect Wiring Diagrams

The basic language of control is the line diagram. Its function is to illustrate quickly and concisely how the control circuit is to perform electrically. The wiring diagram of a circuit can be completed by using the circuit's line diagram. Unlike a line diagram, a wiring diagram shows as closely as possible the actual connections and placement of all components in the circuit.

For example, a standard start/stop pushbutton station with memory is shown by its line diagram. The wiring diagram of the same circuit shows all connections of all components, even those that are present but not used in this particular circuit. See Figure B6. For example, the NC start contacts, NO stop contacts, and the NC auxiliary contacts on the starter are included but not used in the circuit.

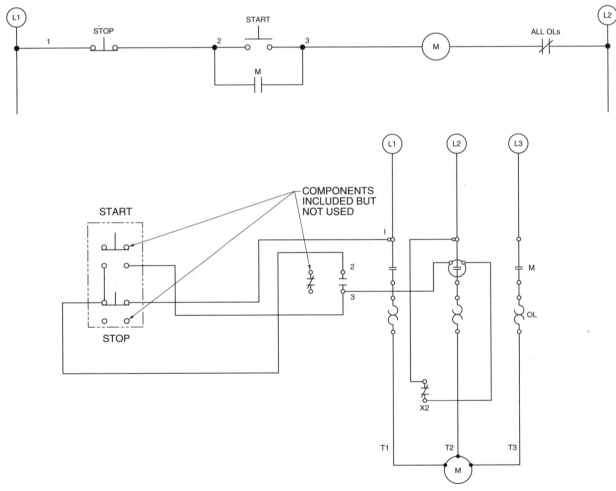

FIGURE B6

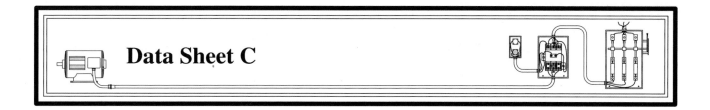

Fluid Power Cylinder Control

The operation of a fluid power cylinder may be controlled by two limit switches and a solenoid-operated valve. See Figure C1. The line diagram illustrates the connections for each piece of electrical equipment.

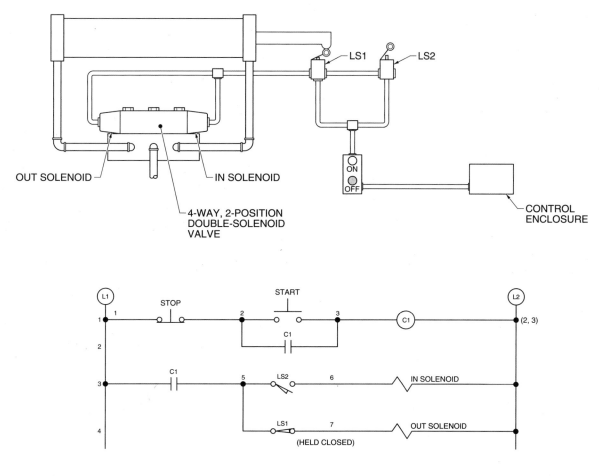

FIGURE C1

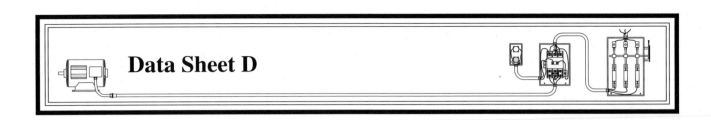

Coding Systems

A coding system may be added to the loads in timing circuits to determine the condition of a load for each sequence of the circuit operation. For example, a code of reset, timing, and timed out may be added to a circuit that uses an ON-delay timer. This coding system is useful in understanding the control circuit and can also be used to design circuits that use timers.

A common ON-delay timer used in a circuit that includes the standard sequence of reset, timing, and timed out has only eight possibilities for any output load. See Figure D1.

TIMER CODING SYSTEM			
Output	Reset	Timing	Timed Out
1	O	O	O
2	O	O	X
3	O	X	O
4	O	X	X
5	X	O	O
6	X	O	X
7	X	X	O
8	X	X	X

O = load de-energized
X = load energized

FIGURE D1

The table may be simplified by eliminating the sequences of OOO and XXX because they do not represent a useful control function. In the case of OOO, the load is de-energized all the time. In the case of XXX, the load is energized all the time. The other six possibilities provide control sequences that are useful and can be accomplished through the contacts normally provided on a timer.

Counting a Coding Table in Contact Arrangements

Timers are normally provided with NO and NC time-delay and instantaneous contacts. By using the timer contacts in various configurations, certain contact arrangements can be associated directly with the remaining six sequence codes in Figure D1.

For example, if an NO time-delay contact is connected in series with a load, the code of OOX results. See Figure D2. Any NO time-delay contact connected in series with a load provides the sequence of OOX. Thus, an electrician can develop a simple table with all six possible combinations that immediately shows which contacts to use and how to connect them for any given sequence. See Figure D3.

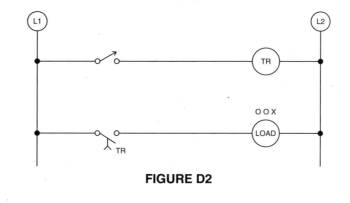

FIGURE D2

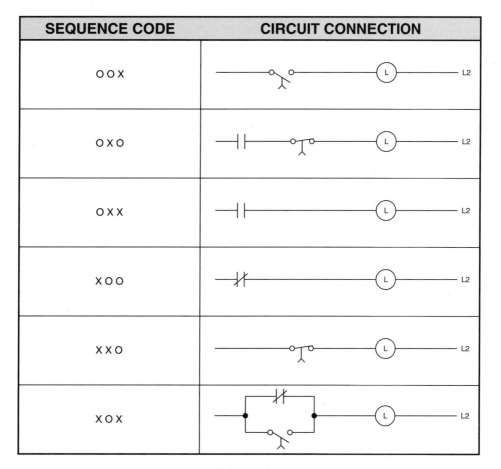

FIGURE D3

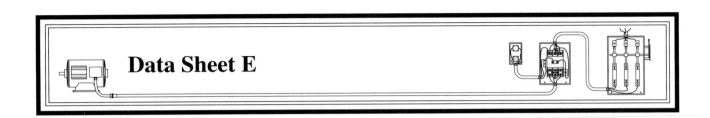

Data Sheet E

Basic Timing Circuit Design

The first step in designing a timing circuit is to list each load according to its requirements in the circuit. The second step is to list the control devices, such as pushbuttons, etc. that are needed. For example, a circuit is required to control the length of time a cake must bake in an oven and indicate by a bell when it must be removed. First, two loads must be controlled, the oven and the bell. The code for the bell is OOX because the bell signals the end of timing. The code for the oven is OXO because the oven bakes for a predetermined length of time. The first part of the circuit is designed from this information and use of a code table. See Figure E1.

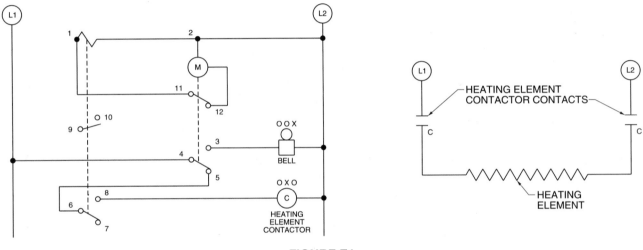

FIGURE E1

The next step in completing the timing circuit is to add the control devices necessary to control the timer and loads. Two controls are required, one to signal the start and end of the baking process and another to control the oven temperature. An ON/OFF switch may be used to signal the start and end of the process, and a temperature switch may be used to control oven temperature. Adding these two controls completes the timing circuit. See Figure E2.

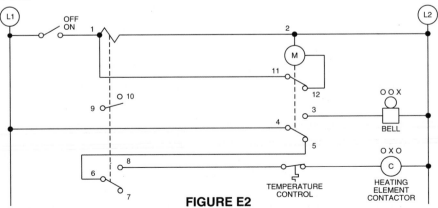

FIGURE E2

Data Sheet F

Level Control Relay

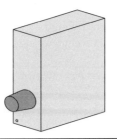

MODEL F

* Level control for conductive liquids
* max.-min control of discharging
* Adjustable sensitivity
* 10 A SPDT output relay
* LED-indication for relay ON
* AC or DC supply voltage

SPECIFICATIONS

Sensitivity
Knob-adjustable sensitivity
with relative scale.

ON from 3.5 kΩ to 25 kΩ
OFF from 8 kΩ to 45 kΩ

Sensor voltage
Max. 24 VAC

Sensor current
Max. 2.5 mA

Connection cable between sensor and amplifier
2- or 3-core plastic cable, normally unscreened.
Cable length: Max. 100 m

Resistance between cores and ground must be
at least 220 kΩ.

In certain cases, use a screened cable between
sensor and amplifier, where cable is placed
parallel to load cables. Screen is connected to
pin 7.

WIRING DIAGRAMS

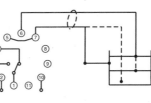

Example 1

Example 2

MODE OF OPERATION

Max. and/or min control for discharging of
conductive liquids.
Relay for control of charging.

Example 1
Level control is connected as max. and
min control (registration of two levels).
Relay operates when liquid reaches max.
electrode (pin 5), provided that min electrode
is in contact with liquid.

Relay releases when min electrode is no
longer in contact with liquid. Pin 7 must be
connected to container. If container consists
of a non-conductive material, an additional
electrode must be used which is connected to
pin 7. This electrode is shown by the dotted line.

Example 2
Level control is connected as max. or min
control (registration of 1 level).
Relay operates when electrode (pin 6) is in
contact with liquid. An additional electrode must
be used if container consists of a non-conductive
material (to be connected to pin 7).

EXAMPLE 1 OPERATIONAL DIAGRAM

Supply voltage	
Max. electrode (pin 5) in liquid	
Min electrode (pin 6) in liquid	
Relay ON	

EXAMPLE 2 OPERATIONAL DIAGRAM

Supply voltage	
Min electrode (pin 6) in liquid	
Relay ON	

Data Sheet G

Reversing Motor Circuits

Electricians are required to wire a variety of AC and DC motors to run in forward and reverse. Wiring the motors can be confusing when the motor wiring diagram is new or unfamiliar. In addition, almost every motor manufacturer provides only one basic diagram. Modifications such as reversing are listed as printed information below the wiring diagram. Thus, the electrician must convert the wiring diagram and written instructions into a circuit that properly reverses the motor.

Basic Wiring Rules

A basic wiring procedure may be modified to develop the proper circuitry to wire any AC or DC motor to run in forward and reverse at any voltage. A wiring diagram can be developed for any motor in just a few minutes after applying the basic wiring procedure to several situations. Basic wiring rules are required before learning the basic wiring procedure. Basic wiring rules include:

- Every hot lead must be switched. The hot lead (ungrounded conductor) must be switched so that the power is never applied to the motor in the OFF position. See Figure G1.
- A motor must be connected for only one voltage. Motors are often designed with multiple windings for different voltages. Always wire the motor for the proper voltage in the specific case.
- A motor must be connected so that it cannot be electrically connected to run simultaneously in forward and reverse directions. Wiring diagrams must provide for the motor to run in only one direction at a time or the motor or power source may be severely damaged.
- A motor must be connected so that it cannot be electrically connected to run simultaneously at different speeds. For a multispeed motor, wiring diagrams must provide for the motor to run at only one speed at a time or the motor or power source may be severely damaged.

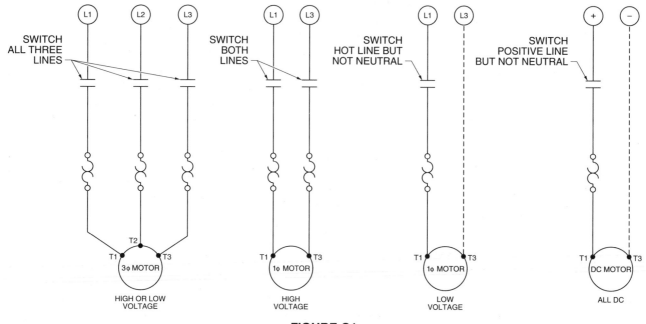

FIGURE G1

Basic Wiring Procedure

Seven steps are necessary for wiring any AC or DC motor to run in forward and reverse.

1. Develop a motor objective. Determine exactly what the wiring diagram is to accomplish. Select only one of the two voltage connections illustrated on the motor. For example, assume that a motor must be connected so that it runs in forward and reverse by means of a magnetic controller.

2. Obtain information from the motor nameplate. The motor nameplate provides information for two voltage supplies. See Figure G2. The nameplate information on a typical 3φ motor shows that it can be wired for 220 V or 480 V. The low-voltage information provided on the motor nameplate is used if the supply voltage required is 220 V.

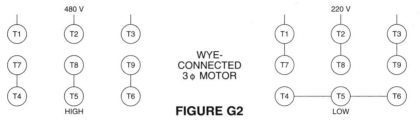

FIGURE G2

3. Make a written diagram of the circuit objective showing exactly how each wire must be connected to accomplish the objective. Put the wiring connections into words. See Figure G3. Follow the manufacturer's wiring diagram and interchange L1 and L3 as specified by NEMA standards.

Forward	Reverse
L1 to T1 and T7	L1 to T3 and T9
L2 to T2 and T8	L2 to T2 and T8
L3 to T3 and T9	L3 to T1 and T7
T4 to T5 to T6	T4 to T5 to T6

FIGURE G3

4. Remove common connections that are not power lines. To simplify the written diagram (or objective), remove any connection that is not the same in forward and reverse, provided it is not the power line. For example, connections T4, T5, and T6 can be removed because they are the same in forward and reverse. Connections L2, T2, and T8 cannot be removed because L2 is a power line and must be switched. See Figure G4.

Forward	Reverse
L1 to T1 and T7	L1 to T3 and T9
L2 to T2 and T8	L2 to T2 and T8
L3 to T3 and T9	L3 to T1 and T7

FIGURE G4

5. Use one name for the remaining common connections. To further simplify the written objective, use one name for any combination of wires that appears in the same direction on both sides of the objective (forward and reverse). For example, T1 and T7 = T1, T2 and T8 = T2, and T3 and T9 = T3. This simplifies the forward and reverse connections. See Figure G5.

Forward	Reverse
L1 to T1	L1 to T3
L2 to T2	L2 to T2
L3 to T3	L3 to T1

FIGURE G5

6. Convert the written objective into a diagram that shows the placement of electric contacts. The wiring diagram may be determined once the simplified written diagram of the objective is developed. A set of electric contacts is needed every place the word "to" appears in the written objective. See Figure G6.

FIGURE G6

7. Draw the wiring diagram. To draw the wiring diagram, only the connections between any two motor or power lines that are the same are required. For example, L1 on the forward side is connected to L1 on the reverse side. Likewise, T1 on the forward side is connected to T1 on the reverse side. The wiring diagram is complete when all the lines are drawn. See Figure G7.

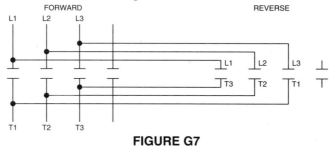

FIGURE G7

Single-Phase Motor Example

Use the nameplate information for a typical 1φ motor that can be wired for 115 V or 230 V to develop the wiring diagram. See Figure G8.

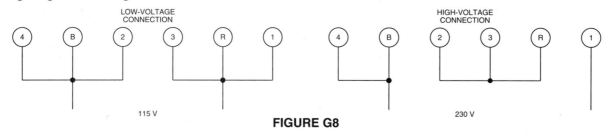

FIGURE G8

1. Develop a motor objective. The objective is to connect a motor to run in forward or reverse by means of a manual controller.
2. Obtain information from the motor nameplate. Assume a supply voltage of 115 V. Obtain the appropriate information from the motor nameplate.
3. Make a written diagram of the objective. See Figure G9. *Note:* The objective has not been simplified. Do not combine steps or simplify the objective too early. Power lines must be switched, therefore, each must be listed separately. The two wires to be interchanged are black and red (B and R). Consequently, they have to be listed separately and should not be connected to any other wire.

Forward	Reverse
B to 4 and 2	R to 4 and 2
R to 3 and 1	B to 3 and 1
L1 to 4 and 2	L1 to 4 and 2
L2 to 3 and 1	L2 to 3 and 1

FIGURE G9

4. Remove common connections that are not power lines. See Figure G10.

Forward	Reverse
B to 4 and 2	R to 4 and 2
R to 3 and 1	B to 3 and 1
L1 to 4 and 2	L1 to 4 and 2

FIGURE G10

5. Use one name for the remaining common connections. See Figure G11.

Forward	Reverse
B to 4	R to 4
R to 3	B to 3
L1 to 4	L1 to 4

FIGURE G11

6. Convert the written objective into a diagram that shows the placement of electrical contacts. See Figure G12.

FIGURE G12

7. Draw the wiring diagram. See Figure G13.

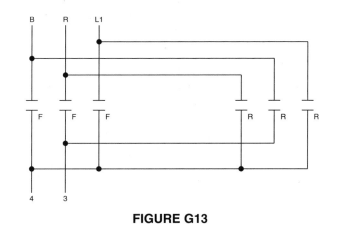

FIGURE G13

Single-Phase, Dual-Voltage, Capacitor-Start Motor Example

Use the nameplate information for a typical 1φ, dual-voltage, capacitor-start motor to develop the wiring diagram for both voltage levels. See Figure G14.

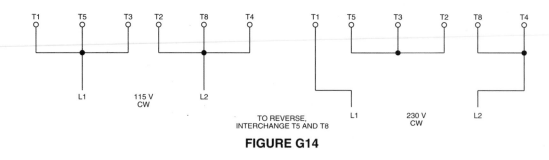

FIGURE G14

1. Develop a motor objective. The objective is to connect a motor to run in forward or reverse by means of a manual controller.
2. Obtain information from the motor nameplate. Two wiring diagrams are required, one for a supply voltage of 115 V and one for a supply voltage of 230 V. The appropriate information for each is obtained from the motor nameplate.
3. Make a written diagram of the objective. A written diagram is required for the 115 V connection and 230 V connection. See Figure G15. T5 and T8 are listed separately because they are to be switched. Also, L1 and L2 are listed separately because they are power lines. No attempt has been made to simplify the objective.

115 V Connection		230 V Connection	
Forward	**Reverse**	**Forward**	**Reverse**
T5 to T1 and T3	T5 to T2 and T4	T5 to T2 and T3	T5 to T4
T8 to T2 and T4	T8 to T1 and T3	T8 to T4	T8 to T2 and T3
L1 to T1 and T3	L1 to T1 and T3	L1 to T1	L1 to T1
L2 to T2 to T4	L2 to T2 to T4	L2 to T4	L2 to T4

FIGURE G15

4. Remove common connections that are not power lines. See Figure G16. The connection L2 to T2 to T4 is removed from the 115 V connection because the connection is the same for both directions and because L2 does not have to be switched. L2 does not have to be switched because it is a neutral wire on 115 V circuits. No connections are removed from the 230 V connection because L1 and L2 are hot wires on 230 V circuits.

115 V Connection		230 V Connection	
Forward	**Reverse**	**Forward**	**Reverse**
T5 to T1 and T3	T5 to T2 and T4	T5 to T2 and T3	T5 to T4
T8 to T2 and T4	T8 to T1 and T3	T8 to T4	T8 to T2 and T3
L1 to T1 and T3	L1 to T1 and T3	L1 to T1	L1 to T1
—	—	L2 to T4	L2 to T4

FIGURE G16

5. Use one name for the remaining common connections. See Figure G17. In the simplified objective for the 115 V connection, T1 and T3 are wired together and named T1. T2 and T4 are wired together and named T2. L2 is to be connected to T2 and T4 and named T2. In the simplified objective for the 230 V connection, T2 and T3 are wired together and named T2.

115 V Connection		230 V Connection	
Forward	**Reverse**	**Forward**	**Reverse**
T5 to T1	T5 to T2	T5 to T2	T5 to T4
T8 to T2	T8 to T1	T8 to T4	T8 to T2
L1 to T1	L1 to T1	L1 to T1	L1 to T1
—	—	L2 to T4	L2 to T4

FIGURE G17

6. Convert the written objective into a diagram that shows the placement of electrical contacts. See Figure G18.

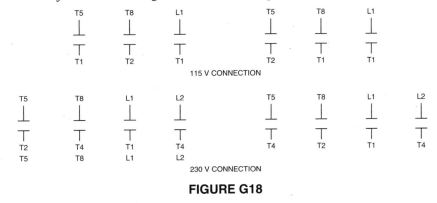

FIGURE G18

7. Draw the wiring diagram. See Figure G19.

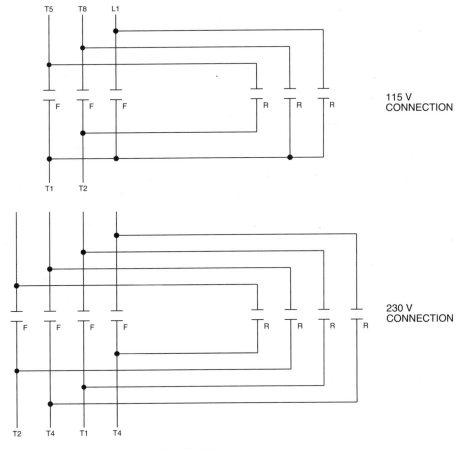

FIGURE G19

The internal connections change within the motor as a result of the high- and low-voltage wiring. See Figure G20.

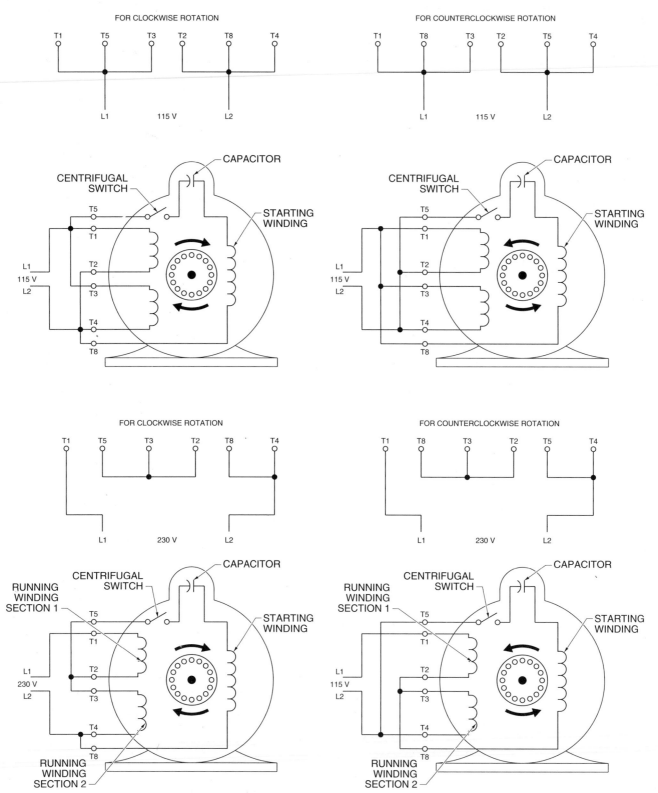

FIGURE G20

Busway Systems

Busway systems are composed of elbows, tees, crosses, feeder ducts, and plug-in ducts. See Figure H1. A sketch is used to help develop a bill of materials for the job.

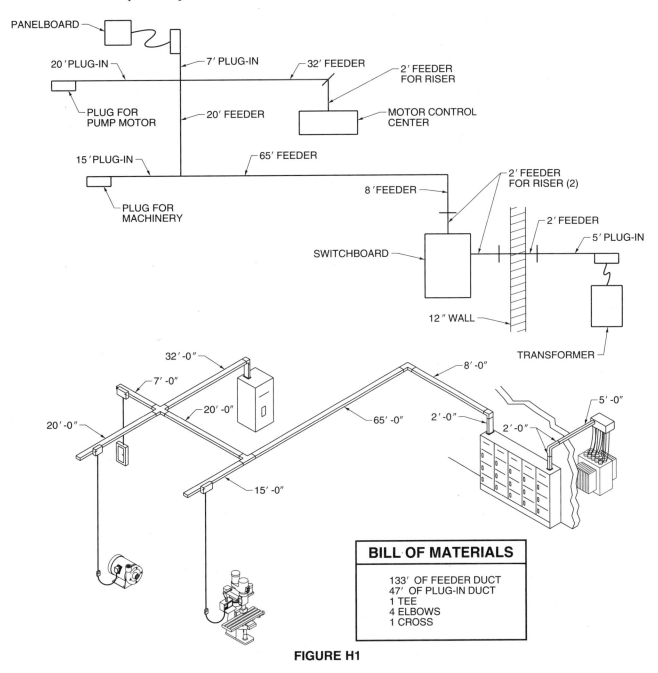

BILL OF MATERIALS

133' OF FEEDER DUCT
47' OF PLUG-IN DUCT
1 TEE
4 ELBOWS
1 CROSS

FIGURE H1

Data Sheet I

Photosensor Relay

MODEL I

* Relay for photosensors with modulated infrared light
* Built-in power supply for transmitter/receiver
* For separate transmitters and receivers with max. ranges 1 m - 100 m
* For combined transmitters and receivers with max. ranges 1 m - 10 m
* Transmitter and receiver connections are short-circuit safe
* 10 A SPDT output relay
* LED-indication for relay ON
* AC or DC supply voltage

SPECIFICATIONS

Response frequency
Max. 10 pulses/sec

Duration of light/darkness
Both: Min 50 ms

Connections for transmitters
Voltage/current = 3.5 VDC / 100 mA

Idle voltage = 5 VDC

Short-circuit current = 500 mA

Connection: Pins 6 and 7
Pin 7 positive
Short-circuit safe

Connections for receivers
Voltage = 12 VDC

Current: Light: 15 mA
Dark: 1 mA – 4 mA

Idle voltage = 12 VDC

Short-circuit current = 75 mA

Connection: Pins 5 and 6
Pin 5 positive
Short-circuit safe

WIRING DIAGRAMS

Note: Standard supply voltage is 115 VAC on pin 2 and 10.

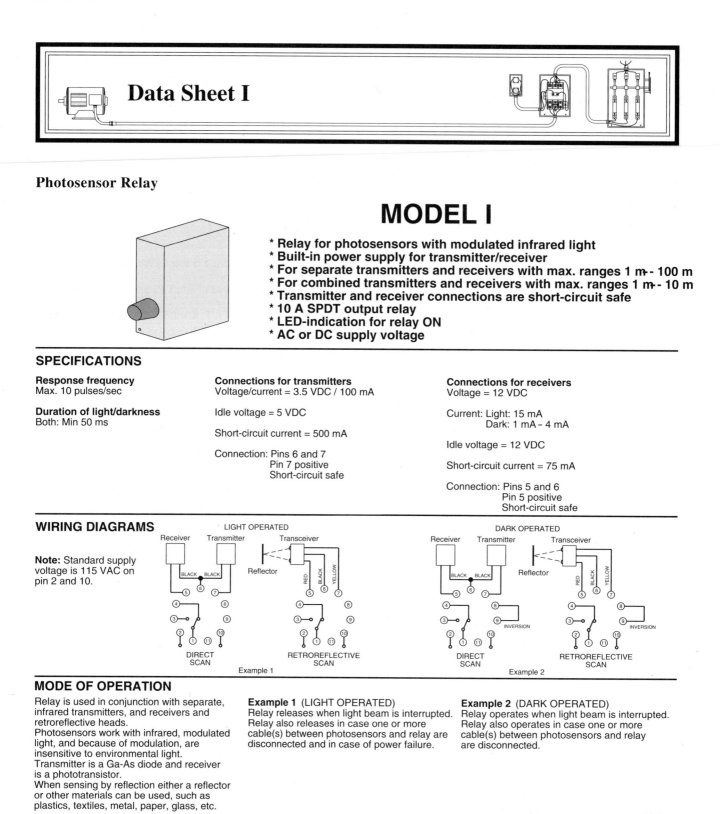

MODE OF OPERATION

Relay is used in conjunction with separate, infrared transmitters, and receivers and retroreflective heads.
Photosensors work with infrared, modulated light, and because of modulation, are insensitive to environmental light.
Transmitter is a Ga-As diode and receiver is a phototransistor.
When sensing by reflection either a reflector or other materials can be used, such as plastics, textiles, metal, paper, glass, etc.

Example 1 (LIGHT OPERATED)
Relay releases when light beam is interrupted. Relay also releases in case one or more cable(s) between photosensors and relay are disconnected and in case of power failure.

Example 2 (DARK OPERATED)
Relay operates when light beam is interrupted. Relay also operates in case one or more cable(s) between photosensors and relay are disconnected.

OPERATIONAL DIAGRAM

Supply voltage	
Light beam interrupted	
Ex 1 Relay ON (light operated)	
Ex 2 Relay ON (dark operated)	

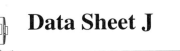

Data Sheet J

Inductive/Capacitive Sensor Relay

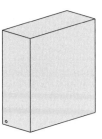

MODEL J

* Relay for inductive and capacitive sensors without amplifier
* Voltage and current limitation in sensor circuit (8 VDC, 8 mA)
* Relay locks in OFF position by cable failures
* 5 A DPDT output relay
* LED-indication for relay ON
* AC or DC supply voltage

SPECIFICATIONS

Sensor voltage
Pins 5 and 6 or 6 and 7:
8 VDC
Pin 6 positive

Sensor current
Activated: < 1 mA
Not activated: > 3 mA

Short-circuit current
Max. 8 mA

Connection cable
Unshielded PVC core. Can be extended if required, maximum resistance = 100Ω .

Sensing range
0.5 mm – 40 mm depending on sensor

Sensing speed
Max. 10 operations/sec

Pulse time
Min 20 ms

Subject of detection
Solid, fluid, or granulated substances

WIRING DIAGRAM

Note: Standard supply voltage is 115 VAC on pin 2 and 10.

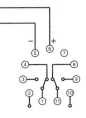

MODE OF OPERATION

Relay operates by activation of sensor.
It releases automatically in case of cable failure.

OPERATIONAL DIAGRAM

| Supply voltage |
| Sensor activated |
| Cable failure |
| Relay ON |

Data Sheet K

Multiple Inductive/Capacitive Sensor Relay

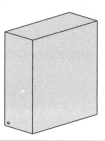

MODEL K

* Bistable relay for 2 inductive or capacitive sensors without amplifier
* Voltage and current limitation in sensor circuits (8 VDC, 8 mA)
* 5 A SPDT output relay
* AC or DC supply voltage

SPECIFICATIONS

Sensor voltage
Pins 5 and 6 or 6 and 7:
8 VDC
Pin 6 positive

Sensor current
Activated: < 1 mA
Not activated: > 3 mA

Short-circuit current
Max. 8 mA

Connection cable
Unshielded PVC core. Can be extended if required, maximum resistance = 100Ω .

Sensing range
0.5 mm – 40 mm, depending on the sensor

Sensing speed
Max. 10 operations/sec

Pulse time
Min 20 ms

Subject of detection
Solid, fluid, or granulated substances.

WIRING DIAGRAM

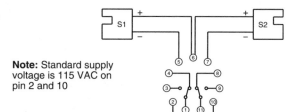

Note: Standard supply voltage is 115 VAC on pin 2 and 10

MODE OF OPERATION

Being a bistable relay, it is used with two proximity sensors.

Relay operates when one sensor (S1) is activated momentarily and then remains operated.

Relay releases when other sensor (S2) is activated momentarily or when supply voltage is interrupted.

If both sensors are activated simultaneously, relay releases or shall not operate respectively. Sensor (S2) has priority.

OPERATIONAL DIAGRAM

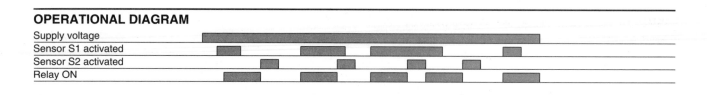

Supply voltage
Sensor S1 activated
Sensor S2 activated
Relay ON

Data Sheet L

Load Guard Relay

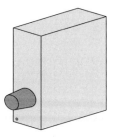

MODEL L

* * Load guard for asynchronous motors and other symmetrical loads
* * Measures phase difference between motor current and voltage
* * Measuring range: Cosφ = 0– 0.9 with current metering transformer
* * Knob – adjustable
* * Delayed function on start
* * 10 A SPDT output relay
* * LED-indication for relay ON
* * AC or DC supply voltage

SPECIFICATIONS

Supply voltage
3 x 220 VAC
Other voltages upon request.

Measuring range
Cosφ = 0 - 0.9
With current measuring transformer.

Hysteresis
10° equaling approx 1 graduation mark.

Adjustment
Knob-adjustable with absolute scale.

Measuring of current phase
Measuring input for connection of current metering transformer:
Pins 8 and 11
Voltage from current metering transformer: 0.1 V_{peak} - 4 V_{peak} . If
current is below 2.5 A conductor may be drawn through central
hole of transformer many times, so that number of turns multiplied
by current consumption is inside current range of transformer.
Transformer should be mounted so current flows from front
towards rear of transformer.

Measuring
Voltage as well as current are
measured on phase connected to pin 5.

Inversion
Output signal can be inverted by
interconnecting pins 9 and 11.

Reaction time
During operation typically 0.5 sec.

WIRING DIAGRAMS

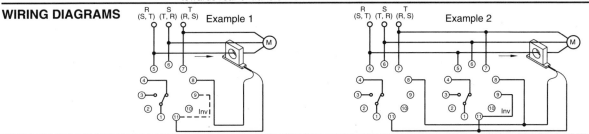

Example 1

Example 2

MODE OF OPERATION

System can be used for monitoring actual
load of asynchronous motors. Measures
angle between motor current and motor voltage,
i.e. phase angle difference. This angle always
exists and its change is almost proportional to
actual motor load (contrary to motor current
solely).
Characteristics of load depend on type of motor
and phase difference. Cosφ depends upon
actual load. It is recommended to adjust cosφ
after practical tests.
Relay contact should be employed as a stop
function in a system with external restart.

Example 1
Relay is connected to a current metering
transformer as well as to a 3φ asynchronous
motor. Relay operates when cosφ is below
set value. At inversion (stippled line) relay
operates when cosφ exceeds set value.

Example 2
By a combination of normal and inverted
functions, relay monitors whether cosφ is
within a set maximum and minimum level
respectively.

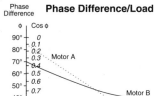

Phase Difference/Load

OPERATIONAL DIAGRAM

Supply voltage

Set value, cosφ
Hysteresis

Relay ON

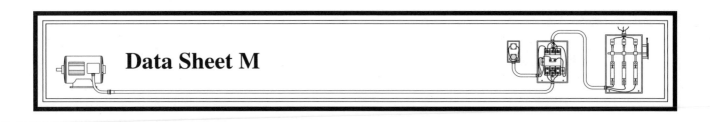

Data Sheet M

Phase Angle Error Relay

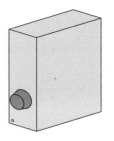

MODEL M

* Relay for phase angle errors and phase breaking
* Metering range for phase angle error: 5°– 15°
* Knob-adjustable phase angle sensitivity
* Operates irrespective of phase sequence
* 10 A SPDT output relay
* LED-indication for relay ON
* Supply voltage is 3-phased metering voltage

SPECIFICATIONS

Inputs
Pins 5, 6, and 7

Metering/Supply voltage
3 x 220 VAC
3 x 380 VAC $\Big\}$ ± 10 %

Frequency
50 Hz or 60 Hz

Reaction time on phase angle error
1 sec is available upon request with
reaction times up to approx 4 sec.

Phase angle sensitivity
5° - 15°± 10%

Amplitude sensitivity
± 30%

Hysteresis
Approx 2°

WIRING DIAGRAM

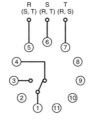

MODE OF OPERATION

Relay meters on its own 3ϕ supply
voltage and controls mutual phase angle.
It operates irrespective of phase sequence
when angle error is less than set value.

In case of interruption of a phase, relay
releases provided that mutual phase angle
error between flawless phases and phase
possibly regenerated by electric motors
connected exceeds set value.

Even if phase angle error does not exceed
set value, relay shall release in case of
phase breaking provided that voltage
regenerated is below 70% of nominal voltage.

OPERATIONAL DIAGRAM

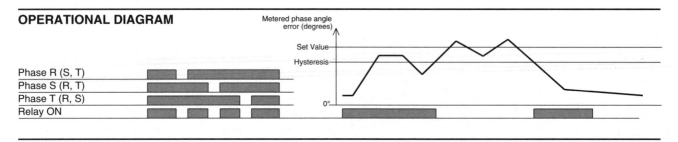

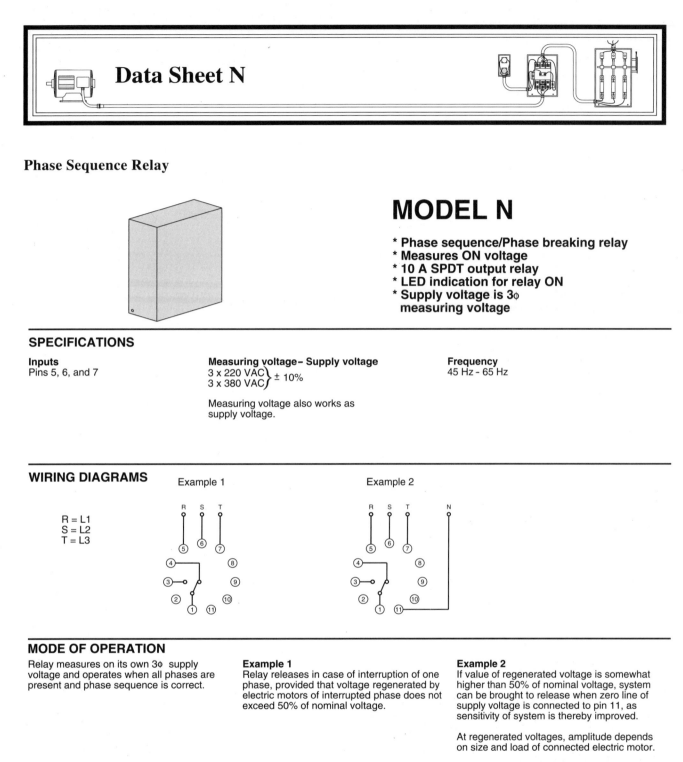

Data Sheet N

Phase Sequence Relay

MODEL N

* Phase sequence/Phase breaking relay
* Measures ON voltage
* 10 A SPDT output relay
* LED indication for relay ON
* Supply voltage is 3φ
  measuring voltage

SPECIFICATIONS

Inputs
Pins 5, 6, and 7

Measuring voltage– Supply voltage
3 x 220 VAC⎱ ± 10%
3 x 380 VAC⎰

Measuring voltage also works as
supply voltage.

Frequency
45 Hz – 65 Hz

WIRING DIAGRAMS

R = L1
S = L2
T = L3

Example 1

Example 2

MODE OF OPERATION

Relay measures on its own 3φ supply
voltage and operates when all phases are
present and phase sequence is correct.

Example 1
Relay releases in case of interruption of one
phase, provided that voltage regenerated by
electric motors of interrupted phase does not
exceed 50% of nominal voltage.

Example 2
If value of regenerated voltage is somewhat
higher than 50% of nominal voltage, system
can be brought to release when zero line of
supply voltage is connected to pin 11, as
sensitivity of system is thereby improved.

At regenerated voltages, amplitude depends
on size and load of connected electric motor.

In practice, value of regenerated voltage can
be near the same as value of supply voltage.

OPERATIONAL DIAGRAM

Phase R Pin 5			S		T		R	
Phase S Pin 6			R		S		S	
Phase T Pin 7			T		R		T	
Relay ON								

Data Sheet O

Current Metering Relay

MODEL O
* **Current metering relay for AC**
* **Metering range: 0.1 A– 500 A (peak) with current metering transformer**
* **Knob-adjustable trip point**
* **Latching at set level possible**
* **10 A SPDT output relay**
* **LED-indication for relay ON**
* **AC or DC supply voltage**

SPECIFICATIONS

Input voltage
Pins 5 and 7: 0.1 V– 4 V
Max. 50 V
Pin 5 positive

Latching
Relay shall latch at set level when pins 8 and 9 are interconnected.

Hysteresis
Approx 10%
Hysteresis can be extended to approx 75% by connecting a resistor between pins 8 and 9. Resistor limits are 1 MΩ and 15 KΩ. Hysteresis increases by decreasing resistance.

AC Measurements 1φ or 3φ
Made in conjunction with one current metering transformer. These transformers deliver an output voltage between 0.1 V and 4 V being proportional with current flowing in a conductor, drawn through center hole of transformer.

WIRING DIAGRAMS

CONNECT PINS 2 AND 10 TO POWER LINES L1 AND L2

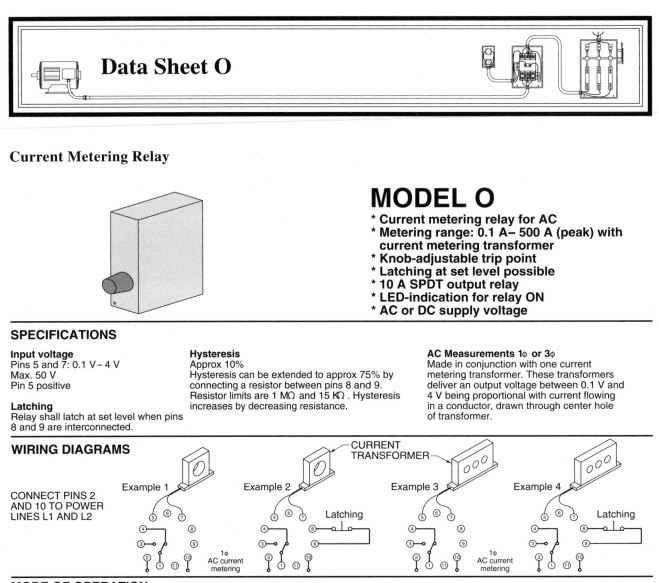

MODE OF OPERATION

Example 1
AC CURRENT METERING (1φ)
Relay operates when current reaches set value. Relay releases when current drops at least 10% below set value (see hysteresis) or by connecting supply voltage.

Example 2
AC CURRENT METERING (1φ) LATCHING
Relay operates when current reaches set value and then latches in operating position.

Relay releases by removing latch, i.e. by opening contact between pins 8 and 9, provided that current has dropped at least 10% below set value (see hysteresis), or by disconnecting supply voltage.

Example 3
AC CURRENT METERING (3φ)
Relay operates when current in any phase reaches set value. Relay releases when current in all 3 phases has dropped at least 10% below set value (see hysteresis) or by

disconnecting supply voltage.

Example 4
AC CURRENT METERING (3φ) LATCHING
Relay operates when current in any phase reaches set value and then latches in operating position. Relay releases by removing latch, i.e. by opening contact between pins 8 and 9, provided that current in all 3 phases has dropped at least 10% below set value (see hysteresis), or by disconnecting supply voltage.

OPERATIONAL DIAGRAMS

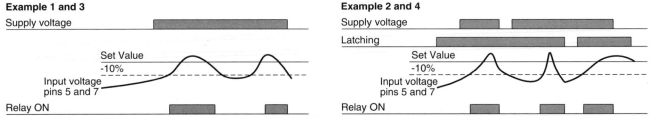

ENCLOSURES				
Type	**Use**	**Service conditions**	**Tests**	**Comments**
1	Indoor	No unusual	Rod entry, rust resistance	
3	Outdoor	Windblown dust, rain, sleet, and ice on enclosure	Rain, external icing, dust, and rust resistance	Do not provide protection against internal condensation or internal icing
3R	Outdoor	Falling rain and ice on enclosure	Rod entry, rain, external icing, and rust resistance	Do not provide protection against dust, internal condensation, or internal icing
4	Indoor/outdoor	Windblown dust and rain, splashing water, hose-directed water, and ice on enclosure	Hosedown, external icing, and rust resistance	Do not provide protection against internal condensation or internal icing
4X	Indoor/outdoor	Corrosion, windblown dust and rain, splashing water, hose-directed water, and ice on enclosure	Hosedown, external icing, and corrosion resistance	Do not provide protection against internal condensation or internal icing
6	Indoor/outdoor	Occasional temporary submersion at a limited depth		
6P	Indoor/outdoor	Prolonged submersion at a limited depth		
7	Indoor locations classified as Class I, Groups A, B, C, or D, as defined in the NEC®	Withstand and contain an internal explosion of specified gases, contain an explosion sufficiently so an explosive gas-air mixture in the atmosphere is not ignited	Explosion, hydrostatic, and temperature	Enclosed heat-generating devices shall not cause external surfaces to reach temperatures capable of igniting explosive gas-air mixtures in the atmosphere
7	Indoor locations classified as Class I, Groups A, B, C, or D, as defined in the NEC®	Withstand and contain an internal explosion of specified gases, contain an explosion sufficiently so an explosive gas-air mixture in the atmosphere is not ignited	Explosion, hydrostatic, and temperature	Enclosed heat-generating devices shall not cause external surfaces to reach temperatures capable of igniting explosive gas-air mixtures in the atmosphere
9	Indoor locations classified as Class II, Groups E or G, as defined in the NEC®	Dust	Dust penetration, temperature, and gasket aging	Enclosed heat-generating devices shall not cause external surfaces to reach temperatures capable of igniting explosive gas-air mixtures in the atmosphere
12	Indoor	Dust, falling dirt, and dripping noncorrosive liquids	Drip, dust, and rust resistance	Do not provide protection against internal condensation
13	Indoor	Dust, spraying water, oil, and noncorrosive coolant	Oil explosion and rust resistance	Do not provide protection against internal condensation

LOCKING WIRING DEVICES

2-POLE, 3-WIRE

WIRING DIAGRAM	NEMA ANSI	RECEPTACLE CONFIGURATION	RATING
	ML2 C73.44		15 A 125 V
	L5-15 C73.42		15 A 125 V
	L5-20 C73.72		20 A 125 V
	L6-15 C73.74		15 A 250 V
	L6-20 C73.75		20 A 250 V
	L6-30 C73.76		30 A 250 V
	L7-15 C73.43		15 A 277 V
	L7-20 C73.77		20 A 277 V
	L8-20 C73.79		20 A 480 V
	L9-20 C73.81		20 A 600 V

3-POLE, 4-WIRE

WIRING DIAGRAM	NEMA ANSI	RECEPTACLE CONFIGURATION	RATING
	L14-20 C73.83		20 A 125/250 V
	L14-30 C73.84		30 A 125/250 V
	L15-20 C73.85		20 A 3φ 250 V
	L15-30 C73.86		30 A 3φ 250 V
	L16-20 C73.87		20 A 3φ 480 V
	L16-30 C73.88		30 A 3φ 480 V
	L17-30 C73.89		30 A 3φ 600 V

3-POLE, 3-WIRE

WIRING DIAGRAM	NEMA ANSI	RECEPTACLE CONFIGURATION	RATING
	ML3 C73.30		15 A 125/250 V
	L10-20 C73.96		20 A 125/250 V
	L10-30 C73.97		30 A 125/250 V
	L11-15 C73.98		15 A 3φ 250 V
	L11-20 C73.99		20 A 3φ 250 V
	L12-20 C73.101		20 A 3φ 480 V
	L12-30 C73.102		30 A 3φ 480 V
	L13-30 C73.103		30 A 3φ 600 V

4-POLE, 4-WIRE

WIRING DIAGRAM	NEMA ANSI	RECEPTACLE CONFIGURATION	RATING
	L18-20 C73.104		20 A 3φ Y 120/208 V
	L18-30 C73.105		30 A 3φ Y 120/208 V
	L19-20 C73.106		20 A 3φ Y 277/480 V
	L20-20 C73.108		20 A 3φ Y 347/600 V

4-POLE, 5-WIRE

WIRING DIAGRAM	NEMA ANSI	RECEPTACLE CONFIGURATION	RATING
	L21-20 C73.90		20 A 3φ Y 120/208 V
	L22-20 C73.92		20 A 3φ Y 277/480 V
	L23-20 C73.94		20 A 3φ Y 347/600 V

INDUSTRIAL ELECTRICAL SYMBOLS . . .

DISCONNECT	CIRCUIT INTERRUPTER	CIRCUIT BREAKER WITH THERMAL OL	CIRCUIT BREAKER WITH MAGNETIC OL	CIRCUIT BREAKER W/ THERMAL AND MAGNETIC OL

LIMIT SWITCHES

NORMALLY OPEN	NORMALLY CLOSED	FOOT SWITCHES	PRESSURE AND VACUUM SWITCHES	LIQUID LEVEL SWITCH	TEMPERATURE-ACTUATED SWITCH	FLOW SWITCH (AIR, WATER, ETC.)
		NO	NO	NO	NO	NO
		NC	NC	NC	NC	NC
HELD CLOSED	HELD OPEN					

SPEED (PLUGGING)	ANTI-PLUG	SYMBOLS FOR STATIC SWITCHING CONTROL DEVICES

STATIC SWITCHING CONTROL IS A METHOD OF SWITCHING ELECTRICAL CIRCUITS WITHOUT USE OF CONTACTS, PRIMARILY BY SOLID-STATE DEVICES. USE SYMBOLS SHOWN IN TABLE AND ENCLOSE THEM IN A DIAMOND.

INPUT COIL   OUTPUT NO   LIMIT SWITCH NO   LIMIT SWITCH NC

SELECTOR

TWO-POSITION	THREE-POSITION	TWO-POSITION SELECTOR PUSHBUTTON

TWO-POSITION

	J	K
A1	X	
A2		X

X-CONTACT CLOSED

THREE-POSITION

	J	K	L
A1	X		
A2			X

X-CONTACT CLOSED

TWO-POSITION SELECTOR PUSHBUTTON

CONTACTS	SELECTOR POSITION			
	A		B	
	BUTTON		BUTTON	
	FREE	DEPRESSED	FREE	DEPRESSED
1-2	X			
3-4		X	X	X

X - CONTACT CLOSED

PUSHBUTTONS

MOMENTARY CONTACT				MAINTAINED CONTACT		ILLUMINATED
SINGLE CIRCUIT	DOUBLE CIRCUIT	MUSHROOM HEAD	WOBBLE STICK	TWO SINGLE CIRCUIT	ONE DOUBLE CIRCUIT	
NO	NO AND NC					
NC						

. . . INDUSTRIAL ELECTRICAL SYMBOLS . . .

CONTACTS

INSTANT OPERATING				TIMED CONTACTS - CONTACT ACTION RETARDED AFTER COIL IS:			
WITH BLOWOUT		WITHOUT BLOWOUT		ENERGIZED		DE-ENERGIZED	
NO	NC	NO	NC	NOTC	NCTO	NOTO	NCTC

OVERLOAD RELAYS

THERMAL	MAGNETIC

SUPPLEMENTARY CONTACT SYMBOLS

SPST NO		SPST NC		SPDT		TERMS
SINGLE BREAK	DOUBLE BREAK	SINGLE BREAK	DOUBLE BREAK	SINGLE BREAK	DOUBLE BREAK	SPST SINGLE-POLE, SINGLE-THROW
						SPDT SINGLE-POLE, DOUBLE-THROW

DPST, 2NO		DPST, 2NC		DPDT		DPST DOUBLE-POLE, SINGLE-THROW
SINGLE BREAK	DOUBLE BREAK	SINGLE BREAK	DOUBLE BREAK	SINGLE BREAK	DOUBLE BREAK	DPDT DOUBLE-POLE, DOUBLE-THROW
						NO NORMALLY OPEN
						NC NORMALLY CLOSED

METER (INSTRUMENT)

INDICATE TYPE BY LETTER	TO INDICATE FUNCTION OF METER OR INSTRUMENT, PLACE SPECIFIED LETTER OR LETTERS WITHIN SYMBOL.			
	AM or A	AMMETER	VA	VOLTMETER
	AH	AMPERE HOUR	VAR	VARMETER
	µA	MICROAMMETER	VARH	VARHOUR METER
	mA	MILLAMMETER	W	WATTMETER
	PF	POWER FACTOR	WH	WATTHOUR METER
	V	VOLTMETER		

PILOT LIGHTS

INDICATE COLOR BY LETTER	
NON PUSH-TO-TEST	PUSH-TO-TEST

INDUCTORS

IRON CORE
AIR CORE

COILS

DUAL-VOLTAGE MAGNET COILS		BLOWOUT COIL
HIGH-VOLTAGE	LOW-VOLTAGE	
LINK	LINKS	
1 2 3 4	1 2 3 4	

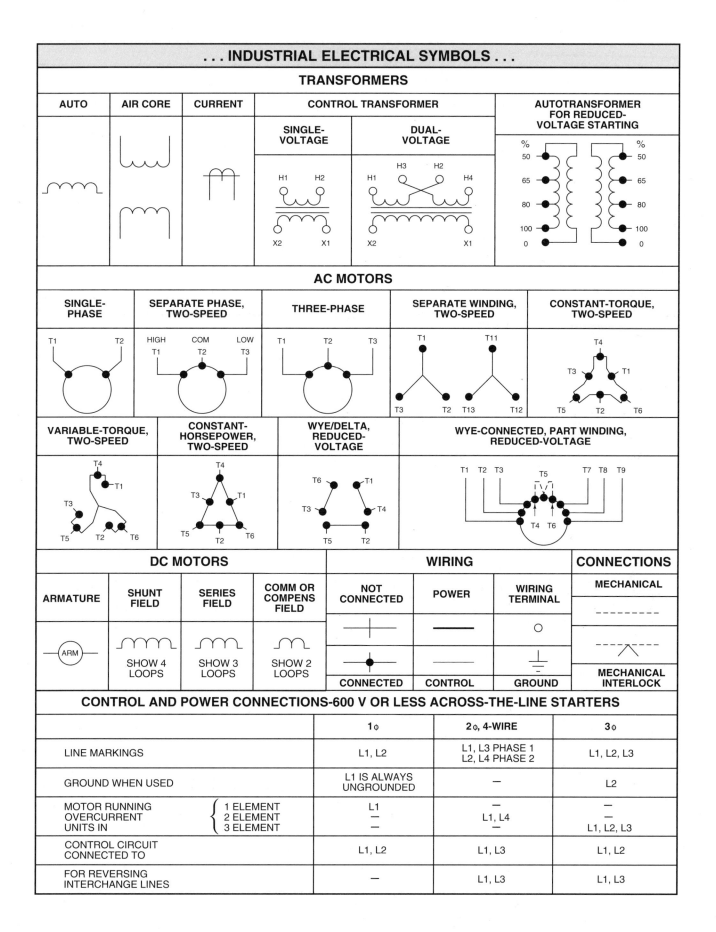

. . . INDUSTRIAL ELECTRICAL SYMBOLS . . .

TRANSFORMERS

AUTO	AIR CORE	CURRENT	CONTROL TRANSFORMER		AUTOTRANSFORMER FOR REDUCED-VOLTAGE STARTING
			SINGLE-VOLTAGE	DUAL-VOLTAGE	

AC MOTORS

SINGLE-PHASE	SEPARATE PHASE, TWO-SPEED	THREE-PHASE	SEPARATE WINDING, TWO-SPEED	CONSTANT-TORQUE, TWO-SPEED

VARIABLE-TORQUE, TWO-SPEED	CONSTANT-HORSEPOWER, TWO-SPEED	WYE/DELTA, REDUCED-VOLTAGE	WYE-CONNECTED, PART WINDING, REDUCED-VOLTAGE

DC MOTORS				WIRING			CONNECTIONS
ARMATURE	SHUNT FIELD	SERIES FIELD	COMM OR COMPENS FIELD	NOT CONNECTED	POWER	WIRING TERMINAL	MECHANICAL
	SHOW 4 LOOPS	SHOW 3 LOOPS	SHOW 2 LOOPS	CONNECTED	CONTROL	GROUND	MECHANICAL INTERLOCK

CONTROL AND POWER CONNECTIONS-600 V OR LESS ACROSS-THE-LINE STARTERS

		1φ	2φ, 4-WIRE	3φ
LINE MARKINGS		L1, L2	L1, L3 PHASE 1 L2, L4 PHASE 2	L1, L2, L3
GROUND WHEN USED		L1 IS ALWAYS UNGROUNDED	—	L2
MOTOR RUNNING OVERCURRENT UNITS IN	1 ELEMENT 2 ELEMENT 3 ELEMENT	L1 — —	— L1, L4 —	— — L1, L2, L3
CONTROL CIRCUIT CONNECTED TO		L1, L2	L1, L3	L1, L2
FOR REVERSING INTERCHANGE LINES		—	L1, L3	L1, L3

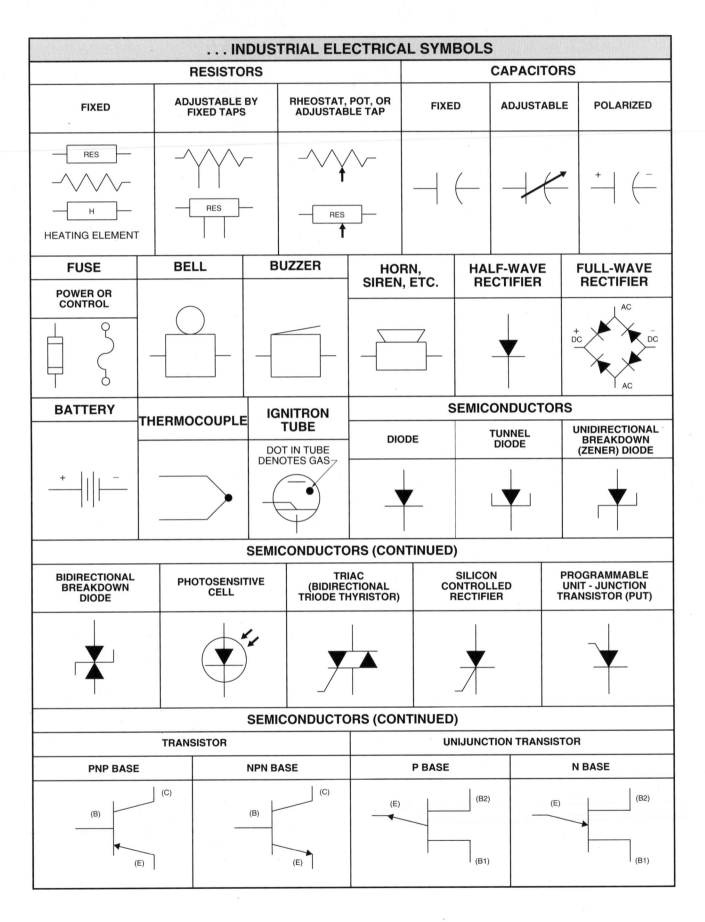

. . . INDUSTRIAL ELECTRICAL SYMBOLS

RESISTORS

FIXED	ADJUSTABLE BY FIXED TAPS	RHEOSTAT, POT, OR ADJUSTABLE TAP

CAPACITORS

FIXED	ADJUSTABLE	POLARIZED

HEATING ELEMENT

FUSE	BELL	BUZZER	HORN, SIREN, ETC.	HALF-WAVE RECTIFIER	FULL-WAVE RECTIFIER

POWER OR CONTROL

BATTERY	THERMOCOUPLE	IGNITRON TUBE	SEMICONDUCTORS		
		DOT IN TUBE DENOTES GAS	DIODE	TUNNEL DIODE	UNIDIRECTIONAL BREAKDOWN (ZENER) DIODE

SEMICONDUCTORS (CONTINUED)

BIDIRECTIONAL BREAKDOWN DIODE	PHOTOSENSITIVE CELL	TRIAC (BIDIRECTIONAL TRIODE THYRISTOR)	SILICON CONTROLLED RECTIFIER	PROGRAMMABLE UNIT - JUNCTION TRANSISTOR (PUT)

SEMICONDUCTORS (CONTINUED)

TRANSISTOR		UNIJUNCTION TRANSISTOR	
PNP BASE	NPN BASE	P BASE	N BASE